THE FALL AND R..SE
OF NUCLEAR POWER
IN BRITAIN

THE FALL AND RISE OF NUCLEAR POWER IN BRITAIN

A History

Simon Taylor

Published by

UIT Cambridge
www.uit.co.uk

PO Box 145, Cambridge CB4 1GQ, England
+44 (0) 1223 302 041

First published in 2016, in England

Front cover photograph by generous permission of and copyright EDF Energy Ltd.
Interior photograph credits shown with the photographs.

Cover design by Andrew Corbett
Interior design by Kate Kirkwood

ISBN: 978 1 906860 31 8 (paperback)
ISBN: 978 1 906860 70 7 (ePub)
ISBN: 978 1 906860 72 1 (pdf)
Also available for Kindle.

10 9 8 7 6 5 4 3 2 1

Contents

Acknowledgements

Many people have kindly helped me with this book, through interviews, suggestions for people to contact, sharing their knowledge and general support. They of course bear no responsibility for what I've written.

I would like to thank: Robert Armour, Peter Atherton, Lord Birt, Jonathan Brearley, John Burnham, Roger Cashmore, Greg Clark, Bent Flyvbjerg, Mark Higson, Lord Howell, Chris Huhne, Julian Huppert, Sue Ion, Professor Sir David King, David MacKay, Geoffrey Norris, Richard Nourse, Willy Rickett, Andrew Shepherd, Adrian Simper, Geoffrey Spence, Tim Stone, Stephen Tindale, Lord Turnbull and Baroness Worthington.

I've learned a lot over the years from my Cambridge colleagues, both in the Energy Policy Research Group, Bill Nuttall and Michael Pollitt, and in the Engineering Department, Cam Middleton, Tony Roulstone and Eugene Shwageraus. I'm very grateful to Shanshan Qiao for taking me to the Daya Bay nuclear station and to John Parsons for introducing me to the nuclear experts of MIT.

Niall Mansfield did an excellent job of editing the manuscript.

And I owe special thanks to Laura-Lucia Richter.

1 / Introduction: Setting the scene

This book tells the story of the rise, fall and second ascendancy of nuclear power in the UK.

It is a story in roughly three acts. The first begins with the birth of civil nuclear power in the 1950s, which brought high hopes for cheap electricity, energy security and export success. These hopes were soon frustrated, but the dream that the UK could build a new, world-leading industry only really faded in the 1980s, when it became clear that the country had cut itself off from the industry-standard reactor types, which were dominated by US designs.

When the Thatcher government, sympathetic to nuclear, privatized the electricity industry, the process revealed how much wishful thinking, if not deceit, there had been in the UK's nuclear sector. British stations, it turned out, were unreliable, costly, and subsequently unsaleable, and were therefore put into a form of long-term managed decline.

In the second act, good management and honest accounting gave the nuclear stations a new lease of life, this time in the private sector. An initially sceptical stock market fell in love with the company created to manage nuclear power, British Energy (BE). However, hubris, financial mistakes and some bad luck led to the government having to rescue a near-bankrupt British Energy in 2002.

Having failed in both the public and private sectors, in the early 2000s the reputation of nuclear power in the UK was at an all-time low. This was when the third act began. As the government of Prime Minister Tony Blair started to examine long-term energy challenges, it found there were two reasons to look again at nuclear power: ageing electricity infrastructure and climate change.

With new optimism that public sector goals could be achieved through private sector efficiency, the government at first tentatively, then enthusiastically promoted nuclear power as something it would allow the private sector to build. The role of the state was to enable: to reform the planning system, ensure proper consultation and remove obstacles. New nuclear would be built without subsidy, and a new generation of safer, more efficient plants would be competitive in a market that factored in the price of damaging carbon emissions.

Then "events" intervened. A global financial crisis made funding these vast projects all but impossible. The new reactor designs turned out to be difficult and costly to build. Significantly, the government's intention of making energy decisions a matter for the market was defeated by the arrival of new, centrally planned targets. These arose from the European renewables directive and then from the UK's own ambitious Climate Change Act of 2008. The act committed the UK to decarbonizing its economy by 80% by 2050, an ambition greater and more specific in scope than that of any other nation.

The implications of the act were not immediately appreciated but over and again they led to a need for more nuclear power. Not only would there need to be investment to replace the ageing British-designed reactors, but also there would have to be a large net addition to capacity, not least because the plan to make fossil fuel stations clean by adding carbon capture and storage (CCS) technology looked increasingly doubtful. Onshore wind farms were economic but unpopular. Offshore wind was acceptable but costly. Solar costs, originally very high, fell rapidly and raised hopes of being fully competitive in a decade or so. Nuclear, on the other hand, while costly and perennially delayed, and while still without a waste disposal solution, would work and would definitely cut carbon emissions.

Britain no longer having indigenous reactor technology meant it had to pick from Japanese, French and American options. The finance too would have to come from abroad – probably from China. The role of the British consumer was simply to pay a very high price for the resulting carbon-free electricity.

It was clear by 2015 that if new nuclear was going to be built it would be with very heavy state involvement: a 35-year power price contract and a state guarantee for the billions of pounds of required financing. The government argued that these policies were justified by the goal of decarbonization. Surprisingly, the European Commission (EC) agreed and approved the deal for the first nuclear station in 20 years, Hinkley Point C (HPC). However, construction still didn't start, mired in disputes between the French and Chinese investors and in contractual arguments with the government.

The outcome of the third act is therefore currently unclear, but it is likely that new nuclear power will happen. It is difficult to see how the government, under pressure to hit the Climate Change Act's carbon budgets, can do without nuclear power, even if the more optimistic forecasts of falling solar costs prove correct. The act has turned into a machine for enforcing investment decisions, just as its proponents intended. What they may not have intended is that it would commit British electricity customers to paying for the most expensive power station in history.

The British nuclear drama is about government policy, about engineering hopes and overambition, and about good intentions and honest mistakes. There is plenty of luck along the way, most of it bad, in the form of external events such as the financial crisis, the Fukushima disaster and the effect of other countries' policies on the UK. Public opinion, once jaded by years of secrecy, false promises of lower costs and scandals at the nuclear reprocessing facility in Sellafield, now appears positive or at least acquiescent.

The latest attempt to make the UK a leading nuclear energy nation is based on a noble ambition to lead the battle against climate change, despite the country accounting for less than 2% of global emissions. Having contributed to the historical growth of carbon dioxide in the atmosphere, the UK certainly has a moral responsibility to help prevent catastrophic damage to the planet. Whether that responsibility is best discharged by spending tens of billions of pounds to make a marginal cut in global emissions is questionable.

If new nuclear eventually goes ahead, it will be at the behest of the UK government's newest friend, the People's Republic of China. A country with which the UK has a controversial history is now essential to the financing and rebuilding of British infrastructure. This could turn out to be a major opportunity for the UK but there is plenty of scope for misunderstandings along the way. What is clear is that the UK's energy policy, one of the most important areas of state responsibility, is no longer entirely in the hands of the British.

Structure of the book

Part I of this volume tells the story of the early hopes for civil nuclear power and the subsequent disappointments. It covers the military origins of nuclear power, the problems of developing commercial

British nuclear stations in the 1960s, and how the abortive attempt to privatize them in 1989 revealed how costly they really were. The second attempt to privatize the newly restructured industry, in 1996, was successful, but in the early 2000s the financial crisis at British Energy discredited the industry once again.

Part II tells how nuclear power moved back on to the policy agenda in the mid-2000s, in response to rising concerns about climate change and UK energy security. Private companies responded enthusiastically to the government's new pro-nuclear approach and the Climate Change Act of 2008 gave extra, though unintended, impetus to nuclear investment.

Part III tells how the new legal carbon emission targets were converted into action. The government made a raft of changes to the planning system, to regulation and to the electricity market, against a remarkable political consensus in favour of nuclear. That consensus survived the Fukushima disaster of 2011.

Part IV concentrates on the leading project of the new nuclear policy, EDF's proposed European pressurized-water reactor (EPR) at Hinkley Point in Somerset. The project became bogged down by arguments among its investors, which include two Chinese nuclear companies, and the rise in construction costs. A former "no subsidy" policy had by 2015 changed to a 35-year price guarantee and other protections for investors.

Part V concludes with thoughts and lessons learned from the history of UK nuclear power.

PART I

Years of hope
and disappointment
(1945-2002)

2

The years of promise and gradual disillusion
(1945-87)

The shadow of the bomb (1945-55)

Britain's pioneering role in civil nuclear power was built on two foundations. The first was Britain's record of nuclear research, which included the world's first artificial nuclear fission, in 1932, at the Cavendish Laboratory in Cambridge. The second was British scientists' involvement in the Manhattan Project, America's scheme to develop a nuclear weapon. Britain was left near-bankrupt at the end of the Second World War, but the desire to remain a "great power" led the government to spend whatever was necessary to build a British atomic bomb.

The British military wanted a copy of the water-moderated reactor the Americans had built at Hanford, in Washington State, because it was proven to work. Plutonium artificially created at Hanford was used in the second of the two bombs dropped on Japan. Codenamed "Fat Man", the bomb targeted Nagasaki and proved to be the more "efficient" weapon.

However, after a period of close military and scientific cooperation during the war, in 1946 the USA abruptly stopped sharing nuclear secrets with its allies, even the UK. The British were forced to find their own way to build a nuclear bomb, and that included creating plutonium.

The British solution was to build a pair of air-cooled reactors at a wartime munitions site called Sellafield near the coastal village of Seascale in what is now Cumbria. The site was secure, relatively remote and had rail and sea access. To avoid confusion with the name of the location of a new nuclear fuel factory at Springfields in Cheshire, the name was changed to Windscale. (The name was changed back to

Sellafield in 1981, after the name Windscale became tainted in the public mind in the 1970s.)

In 1950, three years after the official government decision to build a British atomic bomb, the two large nuclear "piles" at Windscale achieved fission, using graphite as a moderator to control the reaction and creating various new elements, including plutonium. The piles were crude, vertical structures in which the heat generated by the fission was carried away by the air. By early 1952 the plutonium was being separated out from the spent uranium and on 3 October of that year, Britain successfully tested its first nuclear bomb at the Montebello Islands off the north-west coast of Australia, in what was known as Operation Hurricane.

While the urgent priority was to create material for weapons, the government was aware that the heat from nuclear fission could be used to generate electricity. Sir Christopher Hinton, one of a few highly influential nuclear experts in the early years of British nuclear power, was in charge of the construction of Windscale and other key parts of Britain's nuclear infrastructure. In the early 1950s, once the immediate need for plutonium had been met, Hinton recommended building a new, larger reactor that would both manufacture plutonium for the military and provide fission heat to create steam for power generation.

In 1953 Prime Minister Winston Churchill gave the go-ahead to build on the Windscale site what would become the world's first true nuclear power station: Calder Hall.

The design of Calder Hall, while pioneering and highly successful, was different from that of the US reactors and was dictated by the UK's circumstances. Relations with the USA had improved by 1953, but Churchill had failed, in a meeting with President Eisenhower that year, to get much more help, so the UK had to continue to develop its own nuclear technology.

Nuclear reactors all require some way of controlling the fission (the "moderator") and a mechanism for transferring the heat from the fission to a boiler to create steam (a "coolant"). The steam is just like the steam in a conventional power station and any suitable turbine can then be used to turn an alternator, which generates electricity. Nuclear power engineers are fond of saying that nuclear stations are simply very large kettles, using fission as a heat source rather than coal or gas. This is true but is definitely not the whole story.

The UK has access to natural uranium, in which the proportion of the essential fissile isotope of U-235 is 0.7%. The USA had built, at very great cost, factories that enriched this proportion to 80-90% for the uranium-based atomic bomb dropped on Hiroshima. The UK had no such facilities in the mid-1950s, and this stopped the UK following the US water-cooled design, which depended on enriched uranium.

In the early 1950s water-cooled reactors were also thought to be more dangerous, since, if the coolant was for any reason lost, the reactor would overheat and explode. This was tolerable for a reactor sited in a remote part of a large and relatively unpopulated state like Washington, but less helpful for the densely populated UK. One US general explicitly advised the UK against water reactors. It was considered that the only safe site for a water-cooled reactor would be the far north of Scotland, which was not otherwise convenient, not least because there was no need for power there. (Years later two "fast-breeder" reactors were built at Dounreay, on the north coast of Scotland; one duly leaked radioactive material, which is now being cleaned up at great cost)

Another potential reactor design, developed in Canada, used "heavy water", which differs from ordinary "light water" in that each hydrogen atom has not only the normal proton but a neutron as well. The combination of two particles gives this isotope the name "deuterium". (There is also a three-particle isotope of hydrogen called tritium, which is used in thermonuclear bombs.) The extra neutron allows natural uranium to be used as a fuel and the heavy water acts as both moderator and coolant. Heavy water occurs naturally in low concentration in sea water but to isolate it on the scale needed for reactors requires a lot of heavy industrial plant, which the UK also lacked in the 1950s.

Without enriched uranium or heavy water, the UK (like France a few years later) was forced to choose the combination of a graphite moderator and a gas-cooled reactor. At the time there seemed no disadvantage in this design, which had some theoretical advantages over water-cooled reactors. However, it later became clear that the UK had headed down a major dead end.

At this early stage the British electricity industry, although looking forward to nuclear power, didn't contribute to the project. The UK Atomic Energy Authority (UKAEA), a powerful new organization in charge of military and civil nuclear matters, was firmly in control. Chaired by Sir Edwin Plowden, a senior and influential Whitehall figure, the UKAEA was unusually independent and received government funds through a separate "vote", bypassing normal civil service procedures. Treating the UKAEA as an autonomous department reflected the need for rapid decisions.

The UK had no obvious need for new power sources in the late 1940s, as it had apparently abundant supplies of local coal. However, a particularly harsh winter in 1951 forced the country to import coal. The recently nationalized National Coal Board said it couldn't meet the power sector's predicted needs over the next three years, so in 1954 the Conservative government decided to modify some power stations to run on oil. Oil was expensive, however, so new nuclear power began to look appealing. That same year, even though Calder Hall was not yet finished, the UKAEA argued for a major programme to build nuclear power stations that would be on stream by the mid-1960s. A government working party, set up to examine power strategy, enthusiastically backed nuclear power for the future; but it solicited no views from the electricity industry and it overestimated the cost of coal. This would turn out to be the first of several optimistic assessments of the cost of nuclear versus coal-generated electricity.

While it was clear that the economics of nuclear were still inferior to coal, it was thought that the gap would be more than offset by what became known as the "plutonium credit". Plutonium was essential for weapons and had the potential to fuel a later generation of fast-breeder reactors that would have the seemingly magical ability to create their own fuel (by converting otherwise useless material such as U-238 into plutonium). So the plutonium made by Calder Hall was worth something, possibly a great deal, depending on your assumptions about the future and about the number of gas-cooled reactors it could fuel.

In the economic case for new nuclear stations, plutonium was notionally valued at several hundred times the price of gold, though one expert acknowledged that the valuation was rather arbitrary. On this basis, which made nuclear power look economic compared to coal while offering higher security than imports of oil, a 1955 government white paper backed the building of up to 2,000 MW of new nuclear power capacity: roughly four large power stations. The paper described nuclear as "the energy of the future". Geoffrey Lloyd, the Minister of Energy and Power, offered visionary enthusiasm: "Here is new scope for our traditional genius . . . for mixing a small proportion of imported materials with a large proportion of skill, ingenuity and inventiveness. . . . Our nuclear pioneers have now given us a second chance – to lead another industrial revolution in the second half of the twentieth century" (Lloyd, 1955, cited in Hannah, 1982).

The white paper also looked forward to a valuable new export trade at a time when Britain was suffering continual trade deficits. Yet the electricity industry was still barely consulted.

The Calder Hall power station (1956)

Calder Hall was opened officially on 17 October 1956 by the Queen. Generating 65 MW of electrical power from each of its first two reactors, Calder Hall was, in the carefully chosen words of Rab Butler, the government minister who spoke at the ceremony, "the first station anywhere in the world to produce electricity from atomic energy on a full industrial scale" (*The Times*, 1956b). The USA had produced 100 kW of electricity from a nuclear reactor in 1952 at Arco, Idaho. The USSR had produced 5 MW of power from a reactor in Obninsk in June 1954. So Calder Hall was truly original only in its scale. But it was seen as a great British achievement. There was great national pride and talk of a new Elizabethan age.

The Times ran a special supplement on the day of the opening in which Sir Edwin Plowden, Chairman of the UKAEA, looked forward to 15,000 MW of nuclear power by 1975 (compared with the white paper target of just 2,000 MW). He saw Calder Hall not only addressing the UK's perennial threat of fuel shortage, but providing a new base for the country's "true greatness" (Plowden, 1956). Lord Citrine, Chairman of the Central Electricity Authority (CEA), in *The Times* described the British quest for atomic energy: "We embarked on a full programme [of atomic energy] with nothing but green fields and grey matter" (Citrine, 1956). Citrine also tempted fate by adding that "The problems of building nuclear power stations will not differ materially from those met in the construction of conventionally fired stations."

However, there was a touch of *The Wizard of Oz* about proceedings. Ed Wallis, the future Chairman of the privatized electricity generator, PowerGen, was among those who saw the ceremony and was inspired to make his career in nuclear energy. However, Wallis found out some years later that when the Queen pressed the switch that sent power to the grid, the 3 m (10') arrow that swept impressively round the dial showing the number of kilowatts generated was actually moved by a man turning a handle (source: private interview). And although *The Times* made it quite clear that the Calder Hall reactors were "designed primarily to produce the fissile material, plutonium, and their function as power stations is purely secondary" (*The Times*, 1956a), there was no mention of this in Her Majesty's speech.

With Calder Hall in operation the UKAEA argued for an expansion of the original 2,000 MW programme to 3,200 MW. Leaders of the electricity industry, fearful of government restructuring plans, suppressed their doubts about so hastily backing an as yet unproven technology.

1956 was an eventful year for the UK in other ways. On 26 July the Egyptian President, Gamal Abdul Nasser, had nationalized the Suez Canal, in the latest escalation of a long-running dispute with Britain, the former colonial power, and as part of a wider game of playing off the West against the Soviet Union. Two-thirds of oil imports to Europe flowed through the canal, but of equal importance was the fact that Britain and France regarded the nationalization as an outrageous challenge to their waning imperial authority. In what was later revealed to be a conspiracy, on 29 October 1956 Israel invaded the Egyptian Sinai desert and headed for the canal. The next day Britain and France also invaded under the pretence of keeping the peace between Egypt and Israel. The operation had to be humiliatingly reversed when the United States, which had not been informed of the Anglo-French plan, demanded the European troops leave, threatening economic sanctions.

The Suez crisis not only inflicted a crushing diplomatic defeat on Britain, it also stirred up great concern about future oil supplies. As a result, in March 1957 the nuclear programme was scaled up again, this time from 3,200 MW to 5,000-6,000 MW, despite the electricity industry's continuing doubts.

The consequences of hurrying into plutonium production while in possession of incomplete knowledge became alarmingly clear later that year. On 10 October 1957 one of the two Windscale air-cooled uranium piles caught fire, leading to a serious leak of radiation over the Lake District. The fire was visible but caused little concern. The characteristic secrecy around nuclear activities in those days meant that local people only found out about the leak when a press release was issued the following evening. There was near-panic when milk consumption was banned in a 500 square km (220 square mile) area, affecting 660 farms. This was because the main form of radioactive contamination from the fire was iodine-131. While this isotope has a half-life of only eight days (half of its radioactivity decays within eight days), which meant it fairly quickly became safe, iodine is concentrated in cow's milk; hence the ban.

The fire was caused by a phenomenon known as Wigner energy, the poor understanding of which demonstrated the risks involved in rapidly developing a new technology involving hazardous materials.

For perspective, on the International Nuclear Event Scale the Windscale fire was a level five event ("accident with wider consequences"), as was the Three Mile Island incident in 1979. Chernobyl (1986) and Fukushima (2011) were both rated seven ("major accident" – the highest score). At Windscale the piles had air filters installed on the chimneys, which were added only late in construction and at considerable cost, on the insistence of nuclear scientist Sir John Cockroft. The filters greatly reduced the impact of the accident; had the original design, without the filters, been followed, then large parts of the north-west of England may well have become uninhabitable.

The official report into the fire by Sir William Penney was given to Prime Minister Harold Macmillan on 29 October 1957. Macmillan banned publication. Nearly all copies of the report and even the type used to print it were destroyed amid fears that it would jeopardize nuclear cooperation with the USA following the 4 October Sputnik launch, which had caused widespread alarm at the Soviet lead in rocket science. A shortened version of the report was published in November 1957. The full report only appeared in 1992.

One benefit of the fire was that it led to safety improvements at the Magnox power stations, which were later developed from Calder Hall and which also used graphite cores.

The economic case deteriorates (1956-65)

In the years following the Suez crisis, oil prices first rose but then fell as new oilfields came on stream around the world and larger oil tankers came into service that could steam around the Cape of Good Hope and bypass the Suez Canal. Then, after the years of anxiety about insufficient British coal production, in the 1960s the National Coal Board started to forecast a coal *surplus*. Rises in coal and oil prices looked less likely than previously expected. In addition, uranium from Canada became available for bombs, and the USA started to allow the export of enriched uranium to the UK, making the dubious plutonium credit even more questionable. Nuclear running costs were becoming less attractive relative to coal- or oil-generated electricity. On top of this the already significantly higher capital cost of building nuclear stations rose even further owing to higher interest rates. (Nuclear power stations are more capital intensive than oil, coal or gas-fired stations, which makes their overall costs more sensitive to changes in the cost of paying for that capital. This "cost of capital" is in turn governed by the rate of interest paid on the debt that makes up most of that capital.)

More ominous for nuclear power's prospects, the electricity industry, which was, after all, the nuclear stations' only customer, at least in the UK, began to stand up for itself, finally refusing to acquiesce to decisions taken mainly by the UKAEA. In 1957 a new body with responsibility for running the nationalized electricity industry was created: the Central Electricity Generating Board (CEGB). Chaired by Sir Christopher Hinton, the former "nuclear knight" in charge of Windscale, the CEGB allowed the power sector to lose its inhibitions and start criticizing nuclear. As a former leader in the nuclear industry, Hinton might have been expected to support nuclear power stations uncritically, but that was not his way. He had worked his way up from an apprenticeship at the Great Western Railway at 16 to a first-class degree in Mechanical Engineering at Cambridge. Described both as Britain's most brilliant nuclear engineer and as "a little ruthless in pursuit of efficiency" (Williams (1980) p. 102), Hinton in the CEGB saw things rather differently. The nuclear plans were secretly trimmed to nearer 5,000 MW (from 6,000 MW) and the target date extended to 1966 from 1965. Longer term the CEGB would become a formidable opponent to the UKAEA and one of the most powerful actors on the public sector stage, before it was dissolved in 1989 in the break-up and privatization of the industry.

With the Treasury also concerned about rising spending and the coal surplus, a 1960 white paper accepted that nuclear was 25% more expensive than coal and restated the nuclear targets as 3,000 MW by 1965 and 5,000 MW by 1968. The argument shifted towards energy security rather than any immediate economic cost benefit. Nevertheless, this was the largest nuclear power programme in the world, with six power stations under construction.

The new stations were of the "Magnox" type, with a graphite-moderated reactor cooled by carbon dioxide gas, closely modelled on the Calder Hall design. The name arose from the **mag**nesium **non-ox**idising alloy used to clad the fuel canisters. In what was to be a theme of British nuclear design, each reactor was slightly different – for example, early ones had steel containment vessels but later stations used concrete. This reduced the chance of any economies of scale. The Magnoxes were justified mainly on energy security grounds, an argument that had largely vanished by the early 1960s when both coal and oil were abundant. Construction took 40% longer than planned, and the reported economic cost was artificially understated by not charging the research and development costs, by always assuming lower interest rates than actually applied and by the fanciful plutonium credit. Once the CEGB took ownership it re-estimated the cost of nuclear electricity as up to double that of the new Ferrybridge C coal station being built in West Yorkshire in the early 1960s.

The first Magnox after Calder Hall was Chapelcross, built at Annan in what is now Galloway and Dumfries in Scotland, on the site of a former Royal Air Force base. It was pretty much a copy of Calder Hall and the two were owned by the UKAEA, in line with their main purpose of plutonium production for atomic weapons. They shifted mainly to electricity production from 1964. In 1995 the government confirmed that plutonium production for weapons had stopped.

The nine other Magnox stations were intended more for the production of power than of plutonium, though they did contribute to the UK weapons programme (and to nuclear trade with the USA, in which the UK swapped plutonium for the tritium used in the hydrogen bomb). These stations were larger than the Calder Hall/Chapelcross design. Eight (Berkeley, Bradwell, Dungeness A, Hinkley Point A, Oldbury, Sizewell A, Trawsfynydd and Wylfa) were owned by the CEGB, which was in charge of electricity in England and Wales, and one (Hunterston A on the Ayrshire coast south of Glasgow) by the South of Scotland Electricity Board (SSEB).

The Magnoxes performed reliably and safely and were routinely referred to as "workhorses" over the next four decades. Calder Hall, commissioned in 1956 with a 20-year design life, didn't close until 2003. In 2009 the decision was made to close Britain's last Magnox, at Wylfa on the island of Anglesey in Wales, at the end of 2015. Two Magnoxes were exported, one to Japan and one to Italy. North Korea constructed a Magnox-type reactor for plutonium production, suspiciously similar to Calder Hall, at Yongbyon, though no payment was made to the UK.

The Magnoxes now stand silent, simultaneously examples of Cold War-era architecture, and symbols of an optimistic past. The remaining fuel is currently being removed from the closed stations, after which they will be left in "care and maintenance" mode while their radioactivity declines to a point several decades in the future when they can be safely dismantled. Compared with the next generation of British reactors they were a great success.

Enter the advanced gas-cooled reactor (1965-70)

While the Magnoxes were built for plutonium manufacture and energy security, their successors, the advanced gas-cooled reactors (AGRs), were justified on purely economic grounds. The AGR programme turned out to be one of the UK's most disastrous industrial decisions, and not all of that judgement is hindsight. The AGR was a UKAEA design, a development of the Magnox that shared its underlying technology. The lack of enriched uranium and heavy water that had forced the gas-cooled reactor choice in the 1950s no longer applied in the early 1960s, but the UKAEA had invested its money, time and prestige in the AGR, which it thought could be a major export success.

The reason the AGRs were such a disaster was because nearly everything about them was flawed. The design was a scaled-up, higher-pressure and supposedly more efficient version of the Magnox; but no full-scale prototype was built. The stations used various new designs that made construction highly problematic. They used new materials in combinations that had not been properly tested. No two stations were exactly alike so, once again, the benefits of modular construction were lost. And four separate consortia of companies were involved in their building, mainly to ensure that the economic benefits were widely shared. But this placed short-term company gains above the benefits of learning by doing, which a smaller number of consortia might have achieved. In sum, the AGRs were built late, hugely over budget and never performed at the level they were designed for. And not a single one was exported.

Sir Christopher Hinton, the Chairman of the CEGB, wrote as early as 1961 about his doubts regarding the AGR, and he seemed more impressed with the proven Canadian CANDU (Canadian Deuterium Uranium-heavy water) reactors. Under him, the CEGB made it clear that they wouldn't order a single AGR until one was definitely shown to work. The prototype AGR built at Windscale, which at 30 MW was much smaller than the full-scale AGR design of 660 MW, reached full power in February 1963.

The British government was naturally concerned that the AGR would lack export markets if its main local customer, the CEGB, didn't buy one. However, in an April 1964 white paper, *The Second Nuclear Programme*, the government projected a further 5,000 MW of nuclear power being available in the first half of the 1970s, with the CEGB free to choose the reactor type. In theory the AGR might not be chosen.

The CEGB duly invited tenders for two 600 MW reactors at Dungeness, next to the existing Magnox station. By now the USA, which had earlier lacked any urgent need for nuclear power in the light of its huge indigenous energy reserves, was aggressively marketing its water-cooled reactors around the world. General Electric (GE) claimed that its boiling-water reactor (BWR – a type of light-water reactor) would be cheaper than coal. The UK was no longer in the technological lead.

The tender result was announced on 25 May 1965: despite the evidence that light-water reactors were cheaper to build, Dungeness B would be an AGR. Shortly before the tender, Sir Christopher Hinton had retired from the CEGB, leading to speculation about the way the decision had been taken. The published report justifying the choice of AGR over US water reactors was unconvincing, lacking in detail, and made doubtful comparisons between proven (BWR) and theoretical (AGR) power stations.

In particular, it assumed that AGRs would be able to refuel "on-load" (ie without being shut down). A crucial part of any power station's economics is the proportion of the time it actually operates each year, as opposed to being idle for maintenance or for refuelling. Most nuclear stations had to be shut down when new nuclear fuel was inserted. Although some essential maintenance could be done during this "outage" it meant an economically costly loss of power production. For an investment with very high fixed costs, meaning that the costs are pretty much the same whether the station is operating or not, this is a big disadvantage. However, the AGR's on-load refuelling didn't work. Early attempts to refuel on-load in the 1980s led to strong vibrations that shattered the fuel assembly. During renewed efforts in the 1990s a fuel rod got stuck and the idea was abandoned.

Assumptions about construction also appeared overly generous to the UK design, which was not yet built, relative to the actual record of real stations in the USA. There were also concerns about the effect of scaling up the Magnox and about the higher operating temperature.

The AGR was designed to operate at a gas temperature of 650°C. (1,200°F) But the magnesium alloy used to clad the fuel would melt at 640°C (1,185°C) so a new alternative had to be found. On top of all of this, critics pointed to concerns about the graphite cracking under the stress of the extra heat, something that turned out to be a major problem in later AGR operations.

The Labour government, having emphasized the "white heat of technology" in its winning election campaign of 1964, had just cancelled the TSR-2 advanced fighter project developed by the British Aircraft Corporation in favour of buying the US company General Dynamics' F-111 fighter instead, a deal which was itself cancelled later as costs rose. Perhaps making a similar choice of US reactor over a UK one would have been politically too difficult to explain.

In a speech that later would become infamous, Power Minister Fred Lee told the House of Commons he was happy to accept the CEGB's choice of the AGR and further waxed: "I am quite sure we have won the jackpot this time. . . . Here we have the greatest breakthrough of all times" (cited in Burn (1978) p. 10).

The official historian of the electricity industry later described the choice of AGR as "one of the major blunders of British industrial policy" (Hannah (1982) p. 285); but this assessment lay in the future. The official view was that the AGR now made nuclear competitive with coal. The Labour government's October 1965 white paper on fuel policy increased the second nuclear programme from 5,000 MW to 8,000 MW.

The Dungeness disaster

Construction on the new AGR at Dungeness B started in January 1966. A later historian of the privatization of the British electricity industry described it as "the single most disastrous engineering project undertaken in Britain" (Henney (1994) p. 131). Among a certain generation of people, Dungeness B is still a byword for failure of construction, design and project management on a heroic scale. The project was beset by delays, strikes and cost overruns.

The first problems arose with the boiler design. Chairman of the UKAEA, Sir William Penney, assured MPs that "It always happens in this kind of work that some emergency seems to arise, there is a great commotion about something, it looks of course as if it was not right, and then everything converges on it and it almost melts away" (House of Commons (1965) p. 125).

But the problems didn't melt away.

In the late 1960s the CEGB was still committed to the AGR, despite the lengthening list of problems, and the government naturally wanted to avoid public criticism that could damage the supposed export prospects of the AGR. That left the National Coal Board (NCB), another major public body, as the only influential opponent of the AGRs. Nuclear had been justified originally by fears of a coal deficit, which had now disappeared, and by arguments that nuclear would be competitive with the new modern coal stations, which it clearly wasn't.

The Head of the NCB, Lord Robens, had been appointed to bring about an orderly shrinkage of the once mighty coal industry, because it was being displaced as a domestic and industrial heating fuel by gas and on the railways by diesel engines and electrification. Employment in the coal industry had already fallen from around 700,000 in 1958 to 400,000 in 1966 (from a peak of 1.2 million in 1920). But Robens wasn't prepared to shrink it more quickly than necessary. He emerged as a major public critic of nuclear power, especially after the CEGB announced in March 1967 that the second AGR would be built, rather provocatively, at Hartlepool, on the edge of the Durham coal field. In July of that year Robens told the annual conference of the National Union of Mineworkers that the AGRs had cost the country £525m more than coal power stations would have (and 28,000 miners' jobs). In a letter to *The Times* in 1968, Robens pointed out that the projected cost of electricity from the AGRs had risen from 0.19p/unit in the white paper of 1965 to 0.24p in 1968, just three years later. The CEGB replied that it had no pro-nuclear bias and was confident that the AGRs would beat coal on costs by the early 1970s. In 1969 it then ordered a third AGR for Heysham in Lancashire.

The continuing problems of the AGR

The next difficulties emerged in 1969 – with the steel lining. The government said in 1969 that the CEGB had assured it that the problems at Dungeness B had no wider implications for the AGR programme. In 1970 the Nuclear Safety Inspector required changes to the design of later AGRs at Hartlepool and Heysham, which led to further changes at Dungeness B. In 1971 the discovery of corrosion in the Magnoxes led to a redesign at Dungeness B. In 1972 new problems with the chrome steel were discovered, leading to costly replacement at Dungeness B. Construction was too advanced at the other two AGRs so their power rating ➡

was cut instead. It was discovered that the use of methane gas in the coolant gas in order to control corrosion was leading to carbon deposition on the fuel pins, reducing performance. In 1972 *The Times* reported that Dungeness B costs had risen to £170m from the original £89m. The same year, CEGB chairman Arthur Hawkins told the Parliamentary Science Committee that the AGR was an "inherently difficult system" and that the CEGB wouldn't be choosing any more.

In 1976 the now Lord Hinton told the House of Lords that the AGR was an "overambitious extrapolation of designs". A classic analysis in 1977 by economist David Henderson calculated that the waste of British resources on AGRs and on the beautifully engineered but economically disastrous Concorde supersonic airliner were together equivalent to 12 years of university research costs, some £16bn in 2015 prices (author's calculation). Henderson also identified a number of systemic reasons for bad British technology policy, especially a lack of pluralism, as exemplified by the centralism recommended in the 1976 Plowden Committee report (*The Structure of the Electricity Supply Industry*), which recommended that the industry "speak with one voice".

By 1978 the CEGB announced that Dungeness B construction costs had reached £344m, of which £150m was inflation. The CEGB hoped the station would start operating in 1980. Director of Projects at CEGB, Mr C.E. Pugh, was quoted in *The Times*: "Yes in hindsight it was foolish to accept [the original design]". Station Project Manager, Mr F.W. Coates, added his analysis of what had gone wrong with the AGRs: "You don't place a contract with a contractor of doubtful viability. You don't place orders for three designs of AGR. You don't extrapolate the data and jump 200°C without extensive pre-site work. And you do enough research work not to find unpleasant surprises halfway through the project" (*The Times*, 1978).

Dungeness B's first reactor was finally commissioned in 1983, the second in 1988. Then the steam pipes were found to be defective and it was closed. The station reached 50% power only in 1991. Weld defects discovered in 1999 forced the whole plant to close again. It reached close to full power for the first time in 2004, 38 years after construction began. Having been so relatively little used during its operating life, the station was given a 10-year life extension in 2005 and later given further extensions. It is currently scheduled to close in 2028, the extensions in part reflecting how relatively little used the station is.

Reactor wars and the death of British reactors (1971-87)

By the early 1970s, with increasing evidence that the AGR was fundamentally flawed, even the CEGB was losing confidence in it, though not in nuclear more broadly. A large coal-mining strike in 1972 and then the huge increase in oil prices following the Yom Kippur war of 1973 meant that nuclear remained attractive in principle on cost and energy security grounds. However, the AGR now had serious competition, both British and foreign. The options for new nuclear stations were: the AGR; the US pressurized-water reactor (PWR); and two new UKAEA designs – the steam-generating heavy-water reactor (SGHWR) and the high-temperature gas reactor (HTR). Another design, the fast-breeder reactor, was seen by all those who believed in nuclear power as a hope for the longer term. The CEGB's criteria for choosing a reactor were: cost, export potential and safety.

The CEGB was keen on either the HTR or the LWR. But in 1972, MPs of the House of Commons Science Committee backed the UKAEA's water-cooled reactor design, the SGHWR, mainly on the grounds of its export prospects. Critics saw the SGHWR as yet another unusual UK design. Supporters saw potential for collaboration with Canada, which also used heavy water. With a new Labour government elected in 1974, there were further debates in parliament. In July the government backed the SGHWR for new stations at Sizewell in Suffolk and Torness on the east Scottish coast. But the sluggish economy reduced electricity demand, leading to the postponement of both.

In 1976 the US company Westinghouse was reported to be offering a PWR for £200m, compared with the estimated costs of the SGHWR of £370m. Labour Energy Secretary Tony Benn was not keen on the SGHWR but was even more opposed to the US PWR. With the Electrical Power Engineers Association reporting that 95% of members opposed the SGHWR, in the autumn of 1976 Benn launched another review of reactor options. The 1977 report rejected the SGHWR. It saw the PWR as a proven design, but being from the USA it offered no export opportunities for the UK. In another British compromise it recommended buying the PWR and keeping on with the AGR. Two new AGRs were duly announced for Torness and Heysham. Built by a different consortium from those building the earlier AGRs, the design was again distinct from its predecessors. Leading British industrialist Sir Arnold Weinstock lamented publicly the opportunity missed by not adopting the PWR in 1974 and collaborating with the French, who had bought PWR technology in the late 1960s and then started the world's largest programme of nuclear construction.

The year 1979 brought Margaret Thatcher's Conservatives to power, which quickly ended the fantasy that the AGR might provide an export business. Any future nuclear stations built would be PWRs. The Conservatives were instinctively pro-nuclear, partly because they were anti-coal. Nigel Lawson, Energy Secretary and later Chancellor of the Exchequer, described nuclear in his 1992 memoirs as "the means of emancipation from Arthur Scargill [President of the National Union of Mineworkers]". But the Conservatives also intended to bring more market forces into energy policy. This strategy, combined with the later plan to privatize the industry, would stop plans for new nuclear power for two decades.

Oil prices rose sharply again after the Iranian Revolution of 1979-80, and the conventional wisdom was that they would stay high. This reinforced the government's determination to build nuclear as a bulwark against coal. In December 1989 David Howell, Thatcher's first Energy Secretary, told MPs that nuclear was a cheaper form of nuclear power generation than any known to man (quoted in Helm, 2004). He announced 10 new PWRs amounting to 15,000 MW of capacity. But this plan was scaled back in 1982 when the severe economic downturn sharply cut industrial power demand, leaving plenty of spare capacity (as much as 73% in southern Scotland).

Nigel Lawson, who took over as Energy Secretary in 1981, wanted diversification (which was code for more PWRs) and market forces, instead of the planning undertaken by the large team of economists and technocrats at the CEGB. He brought in as new head of the CEGB nuclear physicist Dr Walter Marshall, who was not only pro-nuclear but had the politically useful credentials that Labour Energy Secretary Tony Benn had not liked him. Lawson started out naively on nuclear. Early on he accepted a bet with Thatcher that Dungeness B would be on stream within six months; he lost £10, and learned that this was a standing joke among civil servants.

Britain's adoption of the mainstream PWR technology began with the Layfield inquiry into a new PWR at Sizewell, which ran from 11 January 1983 until 11 March 1985, setting cost and length records. The final report ran to 3,000 pages. A leading British energy economist later described the CEGB's role as "not to present balanced evidence, but rather to gather all the arguments in favour and rubbish the opposition" (Helm, 2004). The CEGB case rested on its forecast of future high oil prices, despite its forecasting skills being badly criticized in a Monopolies and Mergers Commission report in 1991. That argument was poorly received by the inquiry so the CEGB then argued that the PWR would be the cheapest way to meet the need for new capacity.

On 5 December 1987 Sir Frank Layfield delivered his conclusion that the PWR would indeed be the cheapest way to meet that need. Oil prices had by then collapsed but the report was accepted by the government. Gas, which would turn out to be the main source of new power generation in the 1990s, was ignored by the inquiry.

Lawson's hope that nuclear would help the Conservatives defeat the mineworkers came to partial fruition with the bitter 1984-85 miners' strike. Nuclear helped keep the lights on, mainly by avoiding summer outages and saving the equivalent of 2.5m tonnes of coal. However, nuclear's role was much less important than the CEGB's use of mothballed oil power stations, which saved 38m tonnes of coal. Ironically, once the miners had been defeated as a force, nuclear's value as an insurance policy was much reduced. Walter Marshall received a peerage and nuclear appeared to have a secure future under the Conservatives, but not with a British reactor design.

Conclusion

The British, pioneers of nuclear reactor design, were left without a champion at the end of the 1980s when a Conservative government appeared set for a substantial new expansion of nuclear capacity, based solely on economic arguments and with as much market involvement as possible. These arguments, when married with private market ownership, were to prove the greatest challenge to nuclear power yet.

REFERENCES

Burn, D. (1978). *Nuclear Power and the Energy Crisis: Politics and the atomic industry*. London: Macmillan for the Trade Policy Research Centre.

Citrine, L. (1956). "In twenty years' time – meeting the growing need for electricity". *The Times*. 17 October.

Hannah, L. (1982). *Engineers, Managers and Politicians: The first fifteen years of nationalised electricity supply in Britain*. London: Macmillan.

Helm, D. (2004). *Energy, the State, and the Market: British energy policy since 1979*. Oxford: OUP.

Henderson, P. D. (1977). "Two British errors: their probable size and some possible lessons". *Oxford Economic Papers* 29 (2): 159-205.

Henney, A. (1994). *A Study of the Privatisation of the Electricity Supply Industry in England & Wales*. London: Energy Economic Engineering.

House of Commons (1965). *United Kingdom Nuclear Reactor Programme Evidence*. Select Committee on Science and Technology. London: HMSO.

Lloyd, G. (1955). "Speech to Press Conference", 15 February 1955, cited in Hannah (1982).

Plowden, E. (1956). "The second industrial revolution – nuclear power the basis". *The Times*. 17 October.

The Times (1956a). "Faith and decision". 17 October.

The Times (1956b). "The Queen opens the first nuclear power plant". 18 October.

The Times (1978). "Saga of delays at nuclear power station". 26 May.

Williams, R. (1980). *The Nuclear Power Decisions: British policies, 1953-78*. London: Croom Helm.

3
Privatization and failure in the private sector
(1988-2002)

Privatization exposes nuclear true costs

When they were first in government, the Conservatives, led by Margaret Thatcher, had limited privatization ambitions; their main aim was to return to the private sector corporations such as Cable & Wireless and British Aerospace, whose presence in the state sector they saw as unjustified.

With these privatizations successfully complete the government turned to the network utilities: first telecoms, then gas, water and power. Privatization was a powerful political symbol, demonstrating the success of the government's plan to shrink the public sector. It also brought in substantial financial proceeds, which could be used to reduce public debt or to fund tax cuts (usually the latter). It allowed the creation of a mass share-ownership society, with millions of ordinary individuals, who previously had never owned shares directly, becoming shareholders and perhaps permanently inclined to vote for the party of private ownership, the Conservatives. Finally, privatization introduced former public sector companies to private sector management, who were better incentivized to run their operations efficiently.

These goals were to some extent in conflict with each other. The maximum financial proceeds were achieved by compromising on competition. Making millions of retail shareholders happy meant underpricing the new shares, leaving other taxpayers with a lower return.

The conflicts were most apparent in the utility industries because these were formed of companies that were network monopolies. Distribution and transmission of electricity and gas and the piping of water are inherent or "natural" monopolies – there is no sense in duplicating this sort of infrastructure, so effective competition is impossible. Privatizing a monopoly left its new, private sector managers free to exploit customers more efficiently than in the public sector, unless regulated. British Telecom at least faced some nascent competition from the new Mercury telecoms company, which was building a new fibre optic national network. British Gas, however, privatized as an integrated single company, showed how maximizing proceeds for the Treasury could conflict with the need for proper restructuring that protected the customer. The British Gas flotation in December 1986 was a success in terms of funds raised and satisfied retail investors. But the company spent the next 10 years fighting with the regulator and the Monopolies Commission, which were trying to address the results of failing to restructure the company before it was privatized.

Gas was a highly important industry, but electricity was critical to almost every aspect of everyday life and to all parts of the economy. Selling the electricity industry was the most ambitious step yet in the Conservatives' privatization strategy. It was also very large: £37bn of assets and 240,000 employees. This time the government wanted to restructure the industry before privatization, accepting that although this might reduce the value of the assets sold, it would bring the benefits of competition and greater efficiency to the whole economy. However, it was this privatization that would expose the financial and operational weaknesses of the nuclear sector of the industry.

Cecil Parkinson, a leading politician in the Thatcher era, was brought back to government to take charge of the privatization. His goal was to introduce competition into those parts of the industry where it was possible, mainly in power generation. This threatened the existing public sector Central Electricity Generating Board (CEGB), which owned and ran the power stations.

According to a former senior manager and later even more senior figure in the privatized power industry, the functions of the CEGB were: i) to prop up the coal industry (for which the CEGB was its biggest customer); ii) to support the British electrical engineering industry (ditto); iii) to support the rail industry (which moved all the coal); and iv) to provide cheap power to ICI (the leading British manufacturing company in the 1970s and 1980s) (source: private interview, 2006). All of these "worthy" goals were accomplished by forcing a captive customer base to pay dearly for the electricity they received.

Parkinson wanted to avoid the mistakes made when British Gas was privatized as an integrated monopoly; but the pressure from Margaret Thatcher to make nuclear part of the privatization meant he needed to create a company big enough to contain nuclear, along with its financial liabilities. On pure competition grounds the CEGB could have been broken into five roughly equally sized generation companies, enough to create a proper market with real competition (although in the 2000s, when the British power industry finally did contain about five companies, there were still doubts about the extent of real competition).

The compromise plan envisaged only two generation companies, one bigger than the other to accommodate the nuclear stations. The privatization advisers argued it wasn't efficient to split nuclear stations up because the decommissioning costs would be a burden, and that was best borne by a single large company. An equally important reason was that the prospective managers of the future private sector generation companies knew that nuclear would be uncompetitive and it was necessary to compensate the larger company for having nuclear by giving it more of the cheaper coal stations. The two companies became known as Big G and Little G in the structure laid out in the February 1988 electricity privatization white paper.

One important virtue of privatization is that shares are sold through a prospectus, a legally binding document that lays out the financial and operational facts of the businesses being offered for sale. This in turn requires advisers and auditors to verify every number or be candid about the uncertainty.

The Chief Executive of Big G, which later became National Power, was John Baker. The formidably intelligent and articulate Baker had presented the CEGB's case for building the Sizewell B nuclear power station at the Layfield inquiry, in the teeth of considerable scepticism. But as the prospective boss of a privatized power generation company he saw nuclear in a very different light. Nuclear might produce low-cost and reliable power (it later transpired that it didn't) but it brought a huge negative dowry in the form of the liability for the future costs of decommissioning the stations, some of which were only a few years from closing.

Even this might have been manageable if the liabilities could have been pinned down with some accuracy. But the auditors could only estimate them as between £8bn and £11bn, hopelessly uncertain for a sale prospectus. The Treasury reportedly thought they might actually be as much as £15bn. The CEGB, in its public sector accounts, had valued them at only £3.7bn.

The problem was worst for the older stations, the Magnoxes, which were so close to the end of their natural lives that they arguably had a *negative* value: the value of the few years of remaining cash flow generation was less than the cost of closing them and cleaning them up. Since that clean-up cost was unavoidable, it still made sense to run them as long as possible, but this was not an inviting financial prospect for new investors.

On 24 July 1989 Parkinson was forced to tell the House of Commons that the Magnoxes were to be withdrawn from the privatization. This was a setback but not yet a major threat to the policy of building new nuclear stations. National Power still publicly backed the building of three more PWRs, and PowerGen (the former Little G) was investigating a US design called the Safe Integral Reactor; PowerGen was doing this together with Rolls-Royce, which built nuclear reactors for submarines and wanted to expand into the power generation market.

Parkinson was then moved to the Transport department. His replacement, John Wakeham, quickly saw that competition and nuclear privatization were unavoidably in conflict. During the autumn of 1989, with the Magnoxes now out of the privatization, attention turned to the AGRs. Their liabilities were less of an issue given that most had at least a decade of expected operating life. It was instead their operating costs and performance record that started to cause concern.

On 8 November 1989 Margaret Thatcher made a speech to the United Nations on the role of nuclear in reducing the risk of global warming. The following day Wakeham pulled nuclear completely out of privatization and stopped all plans for new nuclear stations pending a nuclear review in 1994. Wakeham privately thought that it would be cheaper to stop work on Sizewell B than continue, but that would have been too much of a political defeat. The nuclear industry was shocked.

Why were the AGRs pulled out of the sale? John Baker, Managing Director of the CEGB until it was broken up and he became Chief Executive of National Power, had written a letter in October arguing that the power from a new PWR would cost 6.25p/unit compared with the coal-generated cost of 3.7p/unit. In other words, new PWRs would be hopelessly uneconomic. How did this square with the CEGB's earlier enthusiasm for Sizewell B and with the desire to build a second PWR at Hinkley Point? Baker's letter laid out how the Hinkley Point figure of 2.24p/unit presented by the CEGB to a planning inquiry had to be "adjusted". First, the cost of capital (the cost of funding the construction) was assumed to be 5% in real (inflation-adjusted) terms for the low-risk CEGB, safe in the public sector. However, a private sector National Power would have a cost of capital of 8%. Two years of inflation since the original estimate pushed up the cost further. The nuclear overheads that had previously been buried somehow in the wider CEGB accounts then needed to be added, plus a higher depreciation charge because the plant would be written off over 20 years instead of 40 years, plus contingencies for construction cost overruns and other risk factors.

The CEGB's last ever annual report showed how far the economics of nuclear had moved. The estimated Magnox decommissioning liabilities had doubled to £600m for each station. The cost of reprocessing the spent fuel (ie extracting plutonium for potential use as fuel) had also doubled, to £6bn. When the CEGB was divided into National Power, PowerGen and the new Nuclear Electric (NE), created to own the English nuclear stations that couldn't be sold, those stations had a current cost (inflation-adjusted) value starting at £6.3bn as at March 1988. The value was then adjusted to minus £1.6bn on 1 April 1989, reflecting the latest estimates of operating costs, future lifetime value and decommissioning liabilities.

The government believed the CEGB had hidden the true costs of nuclear over many years. An authoritative study by Alex Henney argued that the CEGB had indeed distorted the costs by: ignoring research and development costs; underestimating operation and maintenance costs by assuming they would be the same as coal (in the USA they were double); ignoring the additional capital spending needed later to keep the stations running; overstating the cost of competing fuels; ignoring much higher overheads for nuclear; using too low a discount rate (the cost of capital); understating fuel reprocessing costs; understating decommissioning costs (the CEGB made no provision at all until 1976); and using historical cost accounting that flattered these capital-intensive projects during a period of high inflation (Henney, 1994).

The reputation of nuclear power in the UK in 1989 was at an all-time low. The promises, made over many years, that nuclear would become economic were revealed as at first hopelessly optimistic and then dishonest. The older stations had worked reliably but the advanced reactors had continually failed to operate properly. The CEGB had the appearance of an organization that had decided nuclear was the future and had fixed and distorted the numbers to make it seem viable. Only the scrutiny of the private sector had brought all of this to light. The civil servants and CEGB economists might delude each other but couldn't hide from the legal scrutiny of a privatization prospectus.

The CEGB's nuclear employees had always felt like the elite. Now they were utterly demoralized and their whole working lives seemed wasted. Their formerly less prestigious colleagues in the coal stations moved into the private sector with free shares and the prospect of new performance bonuses, although many would also be made redundant in the next few years, reflecting the private sector's unwillingness to tolerate both the gross overstaffing and the "gold plating" of the power stations.

Humiliation leads to transformation

The English and Welsh nuclear stations were put into the new public sector company Nuclear Electric plc (NE). The two Scottish stations were put into Scottish Nuclear Limited (SNL). With no new nuclear investment likely for years to come, both companies looked like containment vessels for a declining industry.

However, the creation of these separate nuclear companies, combined with new and imaginative management, had unexpected results. By focusing only on nuclear, with nowhere to hide losses or weaknesses, and with proper financial accounting, the managements of both NE and SNL achieved a remarkable turnaround in British nuclear performance. Both companies brought about dramatic improvements in productivity and costs, and hugely improved the performance of the unpredictable AGRs.

Though SNL Chief Executive Robin Jeffries was never enthusiastic about privatization, by 1995 he had driven the company into profitability. South of the border, NE, run by the combative Bob Hawley, wanted to give the government the option of privatizing the company by making it unambiguously profitable by 1995, without using subsidies or financial tricks (but excluding the loss-making Magnoxes). For NE this meant improving annual profits by £250m in five years, which the company achieved.

For a power station, the most important productivity measure is how often it is producing power, which is captured by the *load factor*. This is particularly important for a nuclear station because most of its costs are fixed – the costs are much the same whether the station is operating or not. This is because fuel represents a very small proportion of the cost of a nuclear station (perhaps 15%), compared with gas or coal stations (50-90%).

Theoretically a power station could run at 100% load, meaning twenty-four hours a day, seven days a week. However, maintenance breaks are necessary, and for nuclear stations there are outages (intervals when the station is shut down) both for refuelling and for regulatory inspections. The key to maximizing output is to manage the outages so that all necessary tasks are run efficiently and as far as possible simultaneously.

In 1990 NE's load factor was a poor 47%. The troublesome Dungeness B AGR achieved a shockingly bad 12%. But by 1995 the average AGR load factor was nearly 70%. This meant output from these power stations rose by a remarkable 65%, though in large part this improvement showed what a poor job the CEGB had been doing. SNL achieved a 38% output rise in its stations (Taylor, 2007).

NE decided the best way to improve was to identify the world's best practice and copy it. That turned out to be the US utility, Florida Power and Light (FPL), the first ever non-Japanese company to win the prestigious Deming Prize for quality, awarded by the Union of Japanese Scientists and Engineers. NE asked FPL how it did it. NE also used benchmarking information from international organizations such as the World Association of Nuclear Operators (WANO), set up after the Chernobyl disaster in 1986. After Chernobyl the nuclear industry worldwide realized it needed to reassure everyone that the disaster was specific to the Soviet Union. This led to a new spirit of openness, sharing of best practices and safety management and a strict process of peer reviews. In the case of Japan's nuclear industry this was less true, which was one factor contributing to the Fukushima disaster of 2011.

FPL's embrace of Japanese-style management practice was part of a late 1980s and early 1990s fashion prompted by the apparently relentless success of Japanese manufacturing companies such as Toyota and Sony. Though the Japanese economy after 1991 headed into two decades of stagnation, the management lessons remained valid. The key was motivating people through a sense of pride in their work, which required focus on quality and good management.

It was not all cuddly, though. Hawley introduced himself in an interview in the company magazine in June 1992 with the statement: "My job is NOT, repeat NOT, as a hatchet man". However, NE was as overstaffed as other parts of the old CEGB. The workforce shrank from 13,000 people in 1990 to 9,500 by 1995. It was a similar story in Scotland, where the headcount fell by 15% and net cash flow increased from £171m in 1990-91 to £353m in 1994-95.

Nuclear versus coal again

The senior management teams of NE and SNL hoped that their radically improved management of the nuclear stations could lead to the government reconsidering the case for new nuclear stations, although they kept their hopes quiet. A review of nuclear policy was due in 1994, which offered this chance. In the meantime, however, they faced a threat from their old rival: the coal industry.

What later became known as the coal crisis began in late 1992. Coal had been in long-term decline since the 1950s, as, first, the railways switched to diesel from coal locomotives and then domestic heating switched to gas. Nuclear had competed with coal since the early 1960s, further displacing part of coal's traditional market. British coal had become uncompetitive with imported coal, and the failure of the miners' strike in 1984-85 meant the industry could be run down without opposition. On top of all that, in the 1990s coal faced a new, double threat: more stringent limits on the emissions from coal power stations (to curb acid rain caused by sulphur and nitrogen oxides from coal combustion); and a wave of new gas-fired power stations (following a decision that natural gas, previously restricted to use in heating and chemicals, could now be burned in power stations).

The privatization of electricity accelerated the shift from coal to gas, as private power companies opted unsentimentally for whatever was cheapest. The coal industry had been given some protection, in the form of five-year coal purchase contracts with the two generators carved out of the CEGB: National Power and PowerGen. These contracts paid the coal industry prices above market rates, which were passed on in the form of higher electricity prices. This was possible because the newly competitive market in household electricity purchasing didn't start until 1998, so customers were captive. There was therefore no opposition to this deal from either the generators or the regional electricity companies that bought their power.

The coal contracts ran out at the end of March 1993. Both National Power and PowerGen made it clear that they would certainly not pay the same prices in future, since they could now import coal at lower prices and that would be their benchmark. Many British mines were unable to compete on that basis. British Coal, still a government-owned corporation, announced in November 1992 that 31 out of the remaining 50 British coal mines would close unless another deal was done.

Curiously, given the coal industry's long, troubled relationship with the Conservative party, and the fact that the bitterness of the 1985 strike defeat was still fresh, it fell to a Conservative government to arrange a partial rescue. Michael Heseltine had become Secretary of State for Trade and Industry in April 1992 but, finding this label insufficiently grand, had resurrected an old job title and styled himself President of the Board of Trade. The President, admittedly from the more interventionist wing of the party, struck a deal that saved some of the coal mines. He, in effect, bribed the generators to buy British coal at inflated prices for another five years by offering them profit margins they couldn't refuse. So long as National Power and PowerGen were assured of being able to sell the power on to the regional electricity companies, they were happy to pay whatever price the government wanted.

For a government that had strenuously argued for energy policy to be left to the market this was a remarkable turnaround, but national sympathy for the miners, no longer a force that could hold the country to ransom, ran deep. The problem was that intervening to protect coal meant displacing some other form of power. The new, independent gas power stations were protected by contracts with the regional electricity companies, so that left only nuclear to bear the cost.

Ed Wallis, the Chief Executive of PowerGen, with support from many MPs, suggested that the old Magnox stations could close early, creating demand for coal power. Wallis had started his career in the nuclear part of the CEGB but now had a clear commercial interest in keeping nuclear power subordinate to coal.

The Trade and Industry Select Committee of MPs scrutinized the whole situation in parallel with an official government inquiry. NE's Bob Hawley produced evidence that the avoidable cost of nuclear power (that is, ignoring the fixed costs that arose whether the stations ran or not) was now only £12/MWh, below the current power price of £23/MWh. Accountants Ernst & Young audited and confirmed the figure, perhaps the first time in British history that nuclear costs commanded full credibility. From a purely economic point of view it made no sense to limit nuclear output when it was now one of the cheapest forms of power.

The idea of early Magnox closure was therefore rejected by the government, which would have had to pay up earlier for the decommissioning, raising considerably the cost of saving a few coal mines. But the government did ask NE to give up contracts for power, to make room for more coal.

Hawley, Chief Executive of NE (SNL was much less affected by the coal deal) was furious. He refused to break a contract that would cost the company money, arguing that it was in breach of his duty to protect the interest of his shareholder. That very same shareholder – the government – summoned Hawley to London to demand that he do as he was told. Hawley's show of principled independence made clear his commitment to driving the company into profit but did little to widen his fan club in Whitehall.

NE suffered some loss of market share for a while but it did benefit from a market price of power that was kept artificially high; this was because one side effect of the coal deal was to entrench the duopoly of National Power and PowerGen in the generation market. Real competition in power generation didn't come until later in the decade, when prices fell sharply, at great cost to the nuclear industry.

From 1993 NE ramped up its attempt to be taken seriously as a candidate for privatization. It began briefing City analysts as if it were a quoted company, helping to educate the financial industry ahead of what it hoped would be a share sale. It presented steadily improving results and managed its virtual investor relations activities with great professionalism. All of this started to build some credibility in the City, and a number of investment banks became interested in helping sell the company. Not only were output and productivity rising, there was good news on decommissioning. The cost of decommissioning the old Magnox at Berkeley on the river Severn in Gloucestershire was now estimated to be £30m lower than before, another first for an industry where costs had usually turned out to be higher than expected.

Hawley and his team were really convincing the financial community that they had a grip on this technology and were running it as a business. The government by contrast was increasingly irritated. It was a novel situation for a public sector company to be urging the government to privatize it. With the reputation of nuclear among politicians still rather low after the fiasco of the failed privatization, the government was in no mood to be pushed. The exasperated Energy Minister, Tim Eggar told the *Financial Times* in 1993 that it was for ministers to decide whether to privatize companies. He nonetheless saw the long-term future of nuclear as in the private sector.

In October 1993 Hawley then risked antagonizing the government still further by putting forward a planning request for another PWR at Sizewell, to be called station C. Since, pending the now delayed government nuclear review, new nuclear was forbidden, this was a symbolic step, taken mainly to keep the prospect of new nuclear in the review rather than ignored completely, as some civil servants would have preferred.

In March 1994 NE announced that the more modern nuclear stations, the AGRs, had achieved an average load factor of 73.9% compared with 73.2% for the fleet of PWRs in the USA. For the first time the UK could claim a world-leading performance in nuclear power.

The momentum continued when Sizewell B started commissioning in November 1994 and was connected to the national grid in February 2005. The UK finally had a standard, world-class, modern nuclear power station.

The steady improvement in nuclear finances, together with support from the electricity regulator, Stephen Littlechild, led the government to conclude in the nuclear review of 1994 that privatization might be a good idea after all. The sale might raise £1.5-2.0bn, enough to help fund tax cuts ahead of the next general election, due by 1997.

The Treasury, critical of the idea of "national champions" in industry, nonetheless had always tried to favour major British banks in privatizations, which brought both profit and prestige. But as a result of the same government's deregulation of the City of London in the mid-1980s, there was only one credible British bank left capable of handling the sale: Barclays. (Previous distinguished British merchant banks Warburg and Kleinwort Benson had been bought by Swiss and German banks, respectively.) The government duly appointed the Barclays investment bank subsidiary, BZW as financial advisers. Around the same time Morgan Stanley, a major US investment bank, published a report intended for investors (but clearly positioning the bank to take part in a sale) stating that private ownership of nuclear power stations existed in many other countries so privatization in the UK was quite feasible.

BZW's advice was cautiously positive and on 9 May 1995 the government announced the creation of a new, privatized company that would combine the operations of NE in England and Wales with those of SNL. However, only the AGRs and the new Sizewell B station would be included, the old Magnoxes being retained in the public sector for their remaining few years of production. The company would have its headquarters in Scotland, to allay fears of another "stitch-up" in which Scottish jobs were sacrificed for English. The new company was called British Energy (BE), a name that pointedly excluded the word "nuclear".

In December 1995 BE announced that it would not be attempting to use the planning consent for a new station at Hinkley Point C (HPC) it had won in the inquiry that had ended in 1990. Now that it was facing privatization and new private shareholders, the economics of new nuclear were suddenly compellingly bad. The company had quoted evidence from the accountants Price Waterhouse in the nuclear review of 1994 that a new PWR at Hinkley Point or Sizewell could achieve a real return on capital of 5-9% but that the private sector would want at least 11%. Now that the company was heading for the private sector, there was no basis for new nuclear stations, except if the government subsidized the gap. Patrick Green of Friends of the Earth (FoE) was jubilant, telling the *Financial Times* "This is the final nail in the nuclear coffin. . . . No more nuclear power stations will be built in the UK" (*Financial Times*, 1995). There seemed every reason to think he was right.

Privatization (1996)

BE's privatization was one of the last of the Conservative era. The government tried to stoke public excitement with a series of advertisements featuring athletes, under the label "a final burst of energy". The publicity was coy, to put it mildly, about the fact this energy was nuclear in origin. It is not clear what the thousands of retail shareholders thought they were buying, but they had made money on most other British government privatizations and there were financial incentives to buy the shares.

In the wholesale or professional investor market it was a very different story. BZW, acting as global share sale coordinator on behalf of the government, faced a very sceptical investor audience. British institutional investors were particularly concerned about the decommissioning liabilities, given the previous record of escalating costs in the public sector.

There was encouraging investor interest from the USA and Germany, however, where private sector nuclear ownership was quite normal. BZW sent its analysts (including the author of this book) on a relentless campaign of investor meetings across North America, Europe and Asia, gradually educating the investment world about the company, its prospects and the unique situation of a wholly nuclear power company operating in an increasingly competitive electricity market. Nuclear companies in most other countries were parts of diversified power companies, or operated in protected regulated markets, or both. BE was unusually risky, not because of its nuclear operations but because of the prospect of real price competition in the market for power.

The government issued a pathfinder (interim) prospectus on 10 June 1996, with a list of BE's selling points: strong cash flow, no plans to diversify away from nuclear, and a steady management team happy to be judged on its record. The company's future strategy was crucial. Investors are often concerned about companies that generate a lot of free cash flow (meaning cash flow in excess of operating and reinvestment needs). This is because overconfident company management teams can be tempted to put this cash flow into projects or industries in which they lack expertise, rather than returning it, in the form of dividends (or share buybacks), to shareholders, who can then make their own decisions about how best to invest it. Hawley, now Chief Executive of BE, repeatedly reassured prospective investors that it would not go in for what he called "diworsification".

On the evening of Wednesday 10 July 1996 news broke of shutdowns at the Hinkley Point B and Hunterston B stations, after cracks in the steam pipes were discovered in a routine outage. The share sale was due to close for institutional investors on Friday 12 July with the shares then priced over the weekend and new trading to start on Monday 15 July. But retail investors had to commit by Wednesday afternoon; to some newspapers the timing looked suspicious. Some institutions cancelled their bid for shares.

In the flotation of a new company on the stock exchange, the sponsoring bank (in this case BZW) tries to maximize demand so that the existing owner (in this case the government) can receive the highest possible price. The period leading up to the final pricing decision consists of institutional investors submitting bids, conditional on the price ultimately achieved. The bank builds up a demand curve – a schedule of indicative interest in any given price. It then uses judgement and computer software to set a price that will make the seller as happy as possible but not so high that there is a risk that some of the new owners feel taken advantage of. The BE deal was not going particularly well even before the news on 10 July and BZW was advising the government that the share price would be towards the low end of its original hopes.

A distinctive feature of privatizations was that the government wanted a large fraction of the shares to go to retail investors: ordinary people who had filled in forms in the press or visited "share shops" set up by high street banks. The original goal had been that the retail tranche would be 30% of the total, but demand from institutional investors was so low that BZW ended up allocating 50% to the retail investors, in effect using them as ballast to keep the flotation from failure.

On Sunday 14 July, when the shares were being priced and the final decisions taken on how many each applicant would get, the new PWR at Sizewell B announced a possible delay in returning to power because of leaks in fuel pins. BE stressed this was nothing serious. But to some investors the timing of the announcement again seemed calculated, coming as it did after they had lost the chance to change their minds. While this was untrue, it demonstrated the fragility of the whole privatization.

The new shares started trading when the stock market opened on the Monday morning, the new owners having been told the night before how many they had received. If BZW had done its job right, there should have been a small rise in the shares, or at least no fall, reflecting the fact that some investors would have been disappointed with the amount they received, relative to what they wanted, and would buy additional shares in the market.

However, the news from Sizewell B the previous day led one major US investor to sell all of its shares on the first day, an almost unprecedented act. Uniquely for a British privatization the shares fell at the start and closed lower at the end of the first day of trading. Tim Eggar, the Energy Minister, visited the BZW trading floor on the first morning of trading, as was traditional for privatizations. Eggar may not have understood everything about share trading but he could tell the difference between red and green figures on the computer screens displaying the BE share price. The BZW market maker had permission from the government to buy and sell shares in order to "stabilize" the market, a normal form of market manipulation that is temporarily permitted in the early days of a new share trading. The market maker managed to buy enough shares to keep the price up during the 20 seconds that Eggar was watching the screen. In fact BZW's stabilization actions led to the repurchasing of 11.5% of the shares on behalf of the government that had just sold them. While this appeared at first a sign of failure, they were resold in the market at a £33m profit in December 1996, by which time the share price had risen.

Boom and bust in the private sector

After this inauspicious start, BE shares went on to become a stock market darling. Shareholders who had bought the shares at privatization in July 1996 at 203p had by May 1999 received 114p in dividends and still held shares worth 500p. The shares reached 733p in January 1999 but were never to regain that level.

The market's more upbeat view of the shares was driven by the factors that the BZW analysts had pointed to. The company was indeed a cash machine, with higher output, lower costs and higher profits. BE made what seemed well-judged and timely investments in the unfashionable US and Canadian nuclear industries, fulfilling its commitment not to diversify away from what it knew. The company delivered cautious but encouraging messages about the future and gained a reputation for both good execution and effective communications with the market, always important for investors.

There were storm clouds gathering, however. The National Power/PowerGen duopoly had sustained high power prices for longer than most people (including BE's management) had expected. Nevertheless, the electricity regulator had been steadily chipping away, forcing the companies to sell plant to new owners, while brand-new gas stations were gradually making the market more competitive.

The erosion of the duopoly's influence, combined with the prospect of a general oversupply of power arising from the new gas stations, led to a fall in the power price, first gradually and then, as the decade went on, with growing speed.

The point of maximum hubris for any company that later gets into trouble is usually easy to spot with the benefit of hindsight. For BE, alarm bells should have started ringing when the cover of the 1998-99 annual report displayed, not power stations, wires or nuclear fuel

rods but "the art of energy". The images, the report told perplexed
shareholders, "produce a piece of work that conveyed a sense of
innovation, creativity and vitality as well as expressing energy in
the environment and how it touches our everyday lives" (British
Energy (1999) p. 1). This is the sort of nonsense that suggests the
management has lost sight of its priorities. The report, published in
May 1999, roughly coincided with the peak output of the AGRs,
which afterwards became increasingly prone to breakdowns.

    The 1999 annual results presentation meeting was triumphant.
The board, buoyed by news of two further power station life-extension
agreements, decided on a "return of value" (ie a special dividend) of
£432m to shareholders. Admittedly it was under considerable pressure
from those shareholders to pay back some of the prodigious cash flow.
But the timing was ill-judged. The company had until that point had a
strong balance sheet with very little debt, which would have provided
a cushion against the prospect, now widely acknowledged, of a coming
fall in power prices. Paying the special dividend while still funding
investments in the USA and acquisitions in the UK meant taking
on a lot of debt. For a stable, safe utility company, high indebtedness

Figure 3.1 Monthly and 12-month moving average wholesale power prices

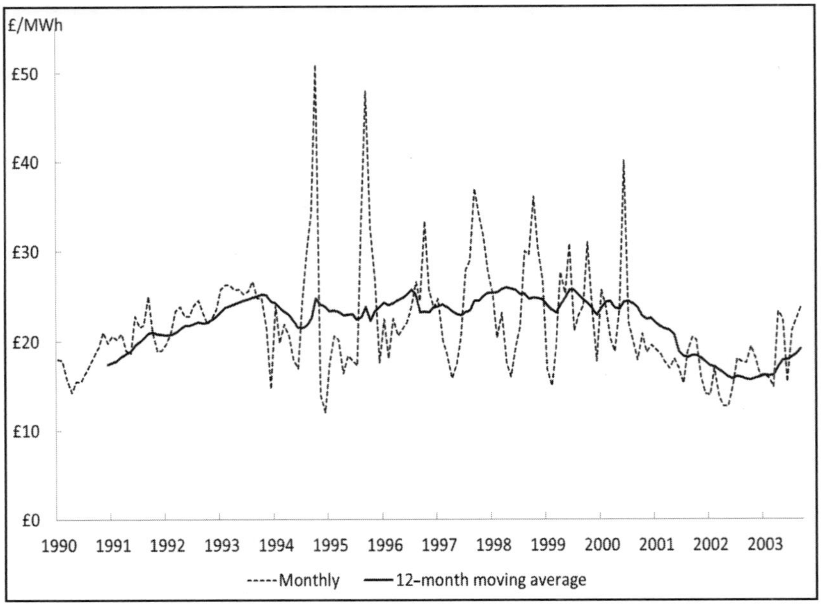

Source: Taylor, 2007

is normal and prudent. But for a company with volatile output and exposure to commodity prices such as electricity, it was far from prudent.

BE's shares fell by 45% during 1999 on fears of lower power prices, which were subsequently proved correct (Figure 3.1). The management couldn't control the power price but it still appeared complacent. Then, owing to a series of breakdowns and failures, the company's electricity output started to fall, casting doubt on whether the former high level of output was sustainable.

UK power prices had been artificially propped up in the first half of the 1990s by the duopoly of the post-CEGB generators, National Power and PowerGen. But high prices encourage new entry and by the end of the decade a lot of new gas-fired power stations had come on stream. The two generating companies could see the writing on the wall and started to sell some of their coal stations to other companies, including BE, which in November 1999 acquired its first non-nuclear station, the 2,000 MW (bigger than most nuclear stations) Eggborough coal station in Yorkshire, for £615m. Was this a case of "diworsification"? Chief Executive Peter Hollins argued that owning a station that could vary its output according to the daily variations in demand, as Eggborough could but nuclear stations couldn't, would allow it to sell power at a higher price. This argument might have had some force, but most of the financial benefit had been transferred to National Power, the purchase price being very near the top of the market.

As National Power and PowerGen lost market share, they also lost pricing power. The market was, at last, becoming more competitive. In any industrial market with overcapacity, producers have an incentive to keep going as long as they cover their variable costs. This was long familiar from industries such as petrochemicals and steel. The UK had the world's first truly competitive wholesale power market, which behaved in the same way as other oversupplied industries.

No generator liked falling prices but BE was worst hit. Nuclear stations aim to run at full capacity because their fuel cost is a very small proportion of the total; so the saving from running at half-capacity is tiny but the loss of income is great. Therefore there was little the company's management could do to offset the falling prices, which directly hit the company's profit and cash flow.

By 10 May 2000 the shares had fallen 80% from their peak. The rating agency Standard & Poor's downgraded the company's credit rating on 12 May from a respectable A- (the highest category is AAA) to BBB+, only two notches above the "junk" or sub-investment grade. Peter Hollins was forced out on June 2001 and Robin Jeffries, the former head of SNL, stepped up to run the whole group. He was respected for his management and engineering skills and the City broadly approved. But it had come to realize that the company was not a safe electric utility; it was just a commodity manufacturer of power, in an oversupplied market.

Jeffries's timing was unlucky. The company's decline was largely unavoidable by late 2000. The combination of worsening AGR reliability, lower prices and the decision to take on more debt left the company with very few options.

By early 2002 power prices had fallen so low that some non-nuclear power stations had been closed. BE could take some comfort from this evidence that the market was beginning to adjust to supply and demand. But it made a £509m provision (a charge against profits) to recognize a fall in the value of the Eggborough coal station it had bought in 1999. In other words, the company had bought for £615m a station that was now worth perhaps £100m.

Financially the company was in a weak and deteriorating position. Net debt to equity, a standard measure of "leverage" or "gearing", was 137% at March 2002 compared with 56% a year before. This was dangerously high for a company that could no longer depend on strong cash flow to service its debt.

The company's financial weakness meant other companies started to question its creditworthiness. Nobody wants to deal with a company that may not be able to honour its debts. This was a particular problem in the British power trading system. BE risked losing its customers if it was thought to be at risk of a sudden crisis.

At this point, the company might still have pulled through, but it was hit by more bad luck. In early 2002 it tried to borrow in the US bond market, issuing what is known as a Yankee bond.[1] But US investors had been shocked by the collapse of the energy company Enron in December 2001, at that time the largest ever US bankruptcy. Nobody wanted to invest in any company related to energy, including electricity, and the bond issue had to be scrapped.

Companies fail for many reasons but the point at which they fail is

---

[1] A Yankee bond is a corporate bond issued in the US market but by a non-US company. Similar bonds in Japan are known as Samurai, in Spain as Matadors, and in the UK as Bulldogs and so on.

essentially the same: They are insolvent when their debts are greater than the value of their assets. A company that is essentially sound but cannot immediately get the cash together to pay its immediate bills is illiquid and can potentially pull through if it can get a temporary credit lifeline. BE was at this point dangerously close to illiquidity but was not yet definitely bust. It still had access to bank loans that might have bought it enough time to keep going until power prices improved, though that could not be predicted with confidence.

During the rest of 2002, however, the company moved relentlessly towards financial insolvency. Its recovery plan increasingly rested on either hoping for prices to recover or getting some form of help from the government. Chairman John Robb had been writing to the government since 2000, arguing that the company's problems were largely not its own fault but a result of government policy. The list included: the fall in power prices caused, or at least made worse by a new power trading system that the government had insisted on imposing; the unfair treatment of nuclear assets in the assessment of business taxes ("rates" in the UK); and the continuing burden of nuclear fuel reprocessing contracts that were a legacy of a policy from the 1970s and imposed a high and more importantly fixed cost on the company.

Patricia Hewitt had become the Secretary of State for Trade and Industry for the Labour government in June 2001. She was firmly in the "New Labour" camp, with a background in civil liberties law and a spell as Head of Research at the consultancy firm Accenture. She had none of the tribal hostility to nuclear that many coal-loving Labour MPs still nursed. But she became increasingly exasperated by the letters from and meetings with BE's senior executives.

One of New Labour's defining features was its embrace of the market. To bail out a former state-owned company would give the impression of a return to the old Labour policy of state intervention and supporting "lame ducks". There might still have been some chance of technical changes to BE's business rates and to the reprocessing contracts; but Hewitt was cautious and expected the company to do everything in its power to save itself first.

The prospect of a deal pretty much vanished when BE announced it would pay a dividend for the financial year to March 2002. The board faced a difficult decision. To cut or scrap the dividend would indicate that the company was in real financial difficulty, which could sabotage commercial relations with other companies and make it even harder to raise new funds. However, the continued dividend sent another signal, one that said things weren't so bad. Hewitt and her advisers drew the conclusion that the company was either not yet in real need or at least wasn't prepared to use its shareholders' funds first; so why should it have any claim on taxpayer funds?

On 10 August 2002 the spot (short-term) power price fell to £10.39/MWh, far below BE's breakeven price of £12. Then the troublesome Dungeness B station had to be shut down again. And at the Scottish Torness station, where one reactor was already closed, a second unit was shut owing to cracks in the giant fan blades that pushed the gas round. BE's shares fell to 69p.

The shares recovered later in August to 85p on hopes of a deal with BNFL (formerly British Nuclear Fuels Limited) to cut the cost of reprocessing nuclear waste and on rumours of a government plan to help. But the BNFL deal fell well short of what BE hoped for. Somewhat perversely, the government-owned BNFL was keen not to be seen as a government department, perhaps because it had hopes of being privatized (these came to nothing – see Chapter 4). BNFL therefore resisted the idea of changing the contracts with BE to recognize the latter's financial difficulty, even though commercially such an arrangement would be entirely reasonable.

BE still had a line of credit with banks giving it the right to borrow up to £615m. That could have provided the liquidity to pay creditors and keep the business afloat for another year or so. But the board was very conscious of its legal liability. It is a criminal offence in the UK to trade while insolvent. A company that knows it can't pay its bills should say so and not carry on doing business with suppliers, which may not be paid. So was BE insolvent?

The board had commissioned a power price forecast from Patricia Hewitt's former employer, Accenture. Consultants are often accused of telling their clients what they want to hear. There was a pessimistic atmosphere at BE and the Accenture forecast was appropriately downbeat. It suggested that prices might fall further and that there was no assurance they would rise again for some time to come. There were good reasons to think the price must rise eventually because the whole power sector was losing money and several other power generators had gone bust, which was helping to cut the overcapacity. Other industries had suffered price collapses. Eventually economic fundamentals would push prices up again, but when? Having asked for a supposedly independent expert opinion, the board felt obliged to rely on it. With the outlook for profits bleak if power prices didn't recover it felt it could not borrow any more without putting itself at legal risk. So the BE board threw away its one financial lifeline.

At a fateful meeting on 5 September the board agreed not to draw on the bank loans. The company put out a statement saying it faced imminent insolvency without immediate financial help. The shares were suspended, as is normal in such dramatic circumstances, to allow shareholders time to digest the news. The shares then reopened on 9 September at 28p. This price reflected some residual hope that a deal might be done to preserve some value for shareholders.

The government was not about to bail out the shareholders. But it felt obliged to provide emergency financial loans to keep the company operating. Normally, when a company enters insolvency, independent accountants take control of it and see how best to realize value from its assets in order to pay back as many of the liabilities as possible. But such a process risked BE having to shut down its stations. This put some 20% of Britain's power at risk. In addition, nuclear stations need to be shut down very carefully; the company's engineers were extremely professional and safety conscious, but whether they would turn up for work if they were no longer being paid was questionable.

The government, now the senior creditor, started to take control of the company, a process that would take some months, because it didn't want the company officially to go into bankruptcy. But there would be no bailout. The shares closed in December 2002 at 5p.

Conclusion

Nuclear power in the UK had failed in the public sector; now it had failed in the private sector. Many people drew the conclusion that nuclear was irremediably uneconomic and unreliable. At the start of 2003 only a naive optimist would have thought that British energy policy was about to swing slowly but firmly back in favour of building new nuclear stations.

REFERENCES

British Energy (1999). *The Art of Energy: Annual report and accounts 1998-99*. May.

Financial Times (1993). "Energy minister attacks Nuclear Electric". 26 November.

Financial Times (1995). "Construction of N-plants axed". 12 December.

Henney, A. (1994). *A Study of the Privatisation of the Electricity Supply Industry in England & Wales*. London: Energy Economic Engineering.

Taylor, S. (2007). *Privatisation and Financial Collapse in the Nuclear Industry: The origins and causes of the British Energy crisis of 2002*. Abingdon: Routledge.

PART II

**Building a new
nuclear policy**
(2002–8)

4

The return of nuclear power to the policy agenda (2002-05)

Why nuclear was unpopular in the early 2000s

The New Labour government elected in 1997 did not have energy, let alone nuclear power, high on its agenda. The electricity industry was important only because it was to pay part of the "windfall tax" on the utility companies (electricity, water and gas), which had been privatized by the Conservatives and were widely regarded as having made too much money for their investors and senior managers, who were pilloried in the press as "fat cats". The industry paid up the £5bn with so little complaint that it seemed to agree that times had indeed been rather too good.

There was no sense of concern about the operations of the electricity network, power supply or security of energy. The collapse of power prices at the end of the 1990s suggested that there was too much power generation capacity, not too little.

Three things conspired to make nuclear unpopular at the start of the new millennium: the Sellafield scandal, the financial collapse of British Energy (BE), and the attitudes of leading Labour ministers.

Sellafield, formerly known as Windscale, was once memorably described in a British newspaper as the "world's nuclear dustbin". It was, indeed, a place where nuclear waste was partially cleaned up, but for the most part the waste was just stored, and not always carefully. A large area of land on the site of a former Second World War munitions store, safely far from large residential areas, Sellafield was a sort of museum of nuclear archaeology. Here were the earliest nuclear "piles" (air-cooled reactors used for manufacturing plutonium, one of which caught fire in 1957), the pioneering Calder Hall Magnox reactor, which closed in March 2003, and a set of waste storage sites

49

(essentially concrete sheds and ponds, some of which had leaked). It also included the MOX facility – a factory for blending plutonium and uranium into mixed oxide fuel. The waste storage and MOX activities were run by a government-owned company, British Nuclear Fuels Limited (BNFL).

Plutonium had been regarded as quite literally priceless in the early 1950s, being an essential ingredient in British atomic weapons before enriched uranium became easily available. Later it was expected to be a source of fuel for fast-breeder reactors, which could then generate their own fuel by converting U-238 into plutonium. But with the end of the Cold War plutonium was in embarrassing oversupply. And the fast-breeder reactor programme was scrapped after a troubled prototype fast reactor at Dounreay in the far north of Scotland was shut in 1994.

The UK found itself with a large inventory of plutonium, which had to be stored securely to prevent it being stolen for use in terrorist bombs. Japan, which had also invested in an abortive fast-breeder programme, found itself in a similar position.

An ingenious solution was to recycle the plutonium into fuel for use in conventional nuclear stations, which used uranium. A small-scale demonstration MOX (mixed oxide fuel) facility was built at Sellafield, and started operation in 1994. It was underpinned financially by contracts with Japan to supply MOX fuel for its nuclear stations. Switzerland and Germany were also customers.

The plant operated without incident until 1999 when the *Guardian* reported that staff at Sellafield had falsified reports on how the MOX fuel was checked before it was dispatched to Japan. Although the fuel had been given an x-ray test, a second, manual test had not taken place and the staff had lied about it. A subsequent Nuclear Installations Inspectorate (NII) report strongly criticized the culture and management at Sellafield. The Japanese customer, Kansei Electric, received financial compensation and a shipment of MOX fuel that was already halfway to Japan had to turn round and return to Sellafield. The Chief Executive of BNFL resigned and the incident wrecked the government's plans to privatize the company (see below).

The scandal revived fears about the secrecy, deceptiveness and sheer incompetence of the old nuclear industry in the UK, concerns that had to a large degree abated since the privatization of BE, with its acknowledged excellent safety culture and open disclosure. But Environment Minister Michael Meacher (see below) and others were unwilling to make much of a distinction between a public sector company that was involved in the nuclear fuel process, and a potential new nuclear power station that would be built by the private sector.

Unfortunately, the most credible and trusted part of the nuclear industry, BE, suffered the financial crisis described in Chapter 3, leading to a request for government funding in September 2002. The financial restructuring of BE dragged on until 2005, when the company was re-listed on the stock exchange with a new management team and the government as its largest shareholder. The message the BE crisis sent to most people was that nuclear power was an economic failure that even private ownership couldn't fix. While this is not necessarily a fair conclusion it did grave damage to the prospect of privately funded and operated new nuclear power stations in the UK.

Anti-nuclear politicians

With the image of nuclear power tainted by Sellafield and BE it was natural that politicians did not see nuclear as having a major role in future energy or environment policy. But the prospects were worsened by the attitudes of the key politicians of the first two Labour governments.

The government ministers with main responsibility for energy and environment were all regarded as anti-nuclear ("totally, utterly and pathologically opposed" was one civil servant's view (source: private interview)). Michael Meacher became Minister of State for the Environment in the New Labour government of 1997 (1997-2001). Meacher was identified as firmly "old Labour", having stood unsuccessfully as the left's candidate for deputy party leader in 1983. Labour party Leader Neil Kinnock once referred to him as "vicar on earth" for the left-winger Tony Benn. Never popular with Tony Blair, perhaps because of his old Labour roots, Meacher nonetheless gained respect for his command of the renewable energy side of policy. But he had the traditional Labour distrust of nuclear, made worse by the scandal at Sellafield, where, in addition, the huge costs of decommissioning were becoming clear in the late 1990s.

At the start of the second Blair government, in 2001, Margaret Beckett became the first Secretary of State for DEFRA – the Department for Environment, Food and Rural Affairs. Beckett was a major figure in old Labour, rooted in working-class politics with a continuing emphasis on state ownership and intervention. This was in contrast to New Labour, led by Tony Blair and Gordon Brown, which accepted much of Margaret Thatcher's criticism of the state and saw state ownership as irrelevant or even damaging to the interests of working people. Beckett had supported left-wing MP Tony Benn in the divisive 1981 election for Labour's Deputy Leader, against the winner, Denis Healey. Elected MP for Derby South in 1983, Beckett supported Neil Kinnock, a modernizer, who successfully stood against Benn for the leadership in 1988. She became a member of Kinnock's shadow cabinet and entered government with Tony Blair in 1997 as President of the Board of Trade (more accurately the Secretary of State for Trade and Industry), with responsibilities that included energy policy. She then became Leader of the House of Commons before in 2001 becoming the first Secretary of State for the newly created DEFRA. In that capacity she had responsibility for policy touching on climate change.

Beckett was regarded by contemporary civil servants as hostile to nuclear power, in keeping with old Labour's traditional opposition to nuclear as the inevitable enemy of coal miners. Years later, she denied she was anti-nuclear and *The Times*, in a 2005 report on the "nuclear cabinet" described her as "persuadable" but "wary", in contrast to the mainly "pro" view of the rest of the cabinet (*The Times*, 2005).

The other important government member with influence over nuclear power was Patricia Hewitt, who was Secretary of State for Trade and Industry from June 2001 to May 2005. Energy policy had been folded into the Department of Trade and Industry (DTI) with the abolition of the separate Department of Energy in 1992. Energy policy under the former Conservative government had been largely left to the market, apart from ad hoc interventions to soothe the euthanasia of the coal industry. Until the early 2000s Labour largely continued the policy of leaving it to the market. So, while the environmental aspect of nuclear fell to Meacher and then Beckett, Patricia Hewitt had oversight of the electricity supply aspect.

Patricia Hewitt was very much part of the modernizing faction known as New Labour and her political background was quite different from Margaret Beckett's or Michael Meacher's. But it was Hewitt who had to deal with BE as it slid towards bankruptcy; and it was she who had to act to keep the company alive when it finally ran out of money in 2002. Understandably, she was left with a poor impression of the British nuclear industry.

Against this very unpromising backdrop, new nuclear power was, nonetheless, creeping back towards the government policy agenda.

The historical background to the revival of interest

Nuclear power's return to government policy in the mid-2000s needs to be examined in the context of the 1990s discussion about climate change. Until the 1990s, the argument for the UK having nuclear power was primarily based on energy security and economics, with a secondary purpose of boosting UK industry. As previous chapters show, only the energy security argument held water, to the extent that nuclear power reduced UK dependence on finite fossil fuel resources. Nuclear was never cost effective and the hopes for exporting UK reactors were repeatedly dashed.

In the 1990s, however, a new argument for nuclear became important. As the evidence grew that the world faced costly and potentially disastrous climate change (initially but inaccurately called "global warming") arising from manmade greenhouse gas emissions, nuclear began to emerge as a source of low-carbon power.

The first British politician to take this threat seriously was Prime Minister Margaret Thatcher. A former research chemist, Thatcher delivered a speech to the United Nations in November 1989 clearly stating the risks of global warming and the need for effective collective action based on more scientific research. She referred to a British scientist working in the Antarctic who saw a risk that climate change could take on a "runaway" quality and become irreversible. She cited the successful international collaboration in the 1980s to stop damage to the ozone layer as a basis for future action. And she called for the Intergovernmental Panel on Climate Change, which the World Meteorological Office and the United Nations Environment Programme had set up in December 1988 in order to provide a report on the matter, to be prolonged after it submitted its first report in 1990.

In her speech, Thatcher endorsed nuclear power "which – despite the attitude of so-called greens – is the most environmentally safe form of energy". But back in the UK her plans for privatization of the electricity industry were exposing the uncompetitive economics of nuclear stations. In a free market, without some form of additional payment for the carbon emissions that nuclear prevents, nuclear was not viable and had to be withdrawn from the privatization.

Thatcher's words were prescient. June 1992 saw 160 countries agree to the United Nations Framework Convention on Climate Change at the "Earth Summit" in Rio. While some parts of the world paid lip service to climate change policy, the European Union started to act.

EU policy centred on the economic concept of an *externality* – something that affects people (it can be good or bad) but which is not captured or priced by the market system. Pollution of any kind is a classic negative externality. A factory pumps polluting effluent into a river, damaging the fishing prospects of people downstream. The factory pays no cost for this pollution and therefore doesn't take it into account in calculating the profitable level of production. Yet damage is done to the fishers. This example points to the typical problem giving rise to an externality: the absence of well-defined ownership or property rights. If somebody owned the river, they could sue the factory for polluting it or come to an arrangement that charged the factory a fee that reflected the damage done (eg the loss of income to the owner of the river from the sale of fishing permits). In that case the externality would be internalized to the market because the factory would now be making output decisions taking into account the pollution cost. The market would then work efficiently without intervention.

The problem of climate change arising from manmade emissions of greenhouse gases is that nobody owns the atmosphere. In the absence of an owner there is nobody to charge polluters for the damage they do. The EU proposed a mechanism that would mimic ownership by charging for the right to emit polluting gases.

There was an encouraging precedent for this scheme in the USA. In 1990 Congress passed the Clean Air Act, to cut acid rain, which is caused by the emission of sulphur dioxide (SO_2) and oxides of nitrogen, which combine with water vapour to create sulphuric and nitric acid, which falls in rain and causes plant damage. The act required quantitative reductions in emissions, with the first deadline being 1995. But rather than just telling companies what to do, the act allowed them to buy and sell the right to emit a quantity of SO_2. This meant that a power generation company that invested in desulphurization equipment and reduced its emissions below the target could sell its surplus to another company that couldn't otherwise meet the target. The system became known as "cap and trade": the act set a cap on emissions, but companies could then trade the rights to emit up to that cap. The benefit of this system, compared with a traditional direct control approach, was that decisions on how best to meet the cap were left to companies, and those that could achieve it at lowest cost had an economic incentive to do so, because they could sell emission rights to those who found it more expensive.

Cap and trade came straight out of economics textbooks and was a great success. Various studies have concluded that it reduced the cost of cutting acid rain emissions by up to 80% compared with the traditional approach of central planning. It was therefore a promising model for the EU when considering emissions caps for greenhouse gases in the late 1990s.

In December 1997 the Kyoto Protocol on Climate Change bound those countries that ratified it (the USA and Australia didn't) to quantitative reductions in six greenhouse gases, chiefly carbon dioxide (CO_2). The protocol specified a range of mechanisms for this, including cap and trade. The EU started working on a possible scheme, leading to a green paper, in February 2000, that laid out a detailed plan for an Emissions Trading System (ETS), which would start in 2005.

The UK was supportive of the Kyoto Protocol and of the EU trading scheme. This was in part because of a fluke of energy policy. The protocol defined 1990 as the base year from which emissions should be reduced (apart from some countries in Eastern Europe, reflecting their turbulent industrial output at the end of the Cold War period). The UK had, quite incidentally, already reduced its CO_2 emissions since 1990 because of a large-scale shift, driven by economics rather than the environment, from coal-fired to gas-fired power stations. This meant the UK had already "banked" a large part of its emission cuts without having to do too much else, at least until the later periods.

The UK already had a policy that appeared to favour low-carbon power production. The Non Fossil Fuel Obligation (NFFO) had been launched in 1990. It took the form of a levy on all electricity produced from fossil fuels, which was distributed to the non-fossil producers. In practice this meant the nuclear industry. There was precious little other non-fossil power in the UK and the NFFO was a rather opaque subsidy to keep the nuclear power sector solvent in the period up to 1996, when BE was finally making a profit without the NFFO subsidy and could be privatized.

The distinction between low-carbon power and renewables is essentially that the former includes nuclear power and the latter doesn't. Confusion between the two was to become a problem in energy policy. With nuclear power struggling in the private sector and nobody taking seriously the prospect of new nuclear, government policy became more directly supportive of other low-carbon sources of power: that is, renewables. In August 2000, to replace the NFFO, the Utilities Act introduced the Renewables Obligation, which came into effect in April 2002 (2005 in Northern Ireland). The obligation set a requirement for electricity suppliers (the companies that buy or produce power and then sell it to end-customers such as companies and households) to source a rising proportion of their total power from renewable sources. It started with 3% in 2002-03 with the intention that it would rise annually to much higher levels.

The mechanism had a cap-and-trade aspect. Owners of renewable energy received a Renewables Obligation Certificate (ROC). Suppliers, needing to show they had fulfilled their obligation, could either invest directly in renewable energy, or buy ROCs from other suppliers at a price determined by the market. So the value of renewable energy was related to how scarce it was. A large-scale investment in renewable energy would reduce the value of the ROCs, and a shortage would drive the price up, which made economic sense. The main types of energy covered were: solar; wind; biomass (burning certain types of plants); landfill gas (methane from decaying waste vegetable matter); and tidal and wave power. In practice the majority of investment has been in wind, with a growing proportion in solar. Tidal has controversial local environmental effects and wave power remain unproven.

Official policy on nuclear leaves the door ajar

Government energy policy in the early 2000s saw little future role for nuclear energy. Against the headwinds of unsympathetic politicians and a financially ailing BE, the abiding theme was that new nuclear was marginal.

The DTI published an energy report in 2001 that mentioned nuclear mainly in the context of the UK's residual involvement in research into Generation IV reactor designs (which are more efficient and safe than earlier types). But the market for such new reactors was international, not domestic:

> The aim of this collaborative international research and development initiative is to identify the technical requirements of, and potential for, innovative nuclear systems post 2030. UK membership of the Forum does not mean a change in nuclear policy in the UK. As with other fuel options it is for generating companies to come forward with proposals for new capacity – but it allows the UK to play its part in ensuring that future reactor designs deployed worldwide achieve high standards of safety, security, waste minimisation and public acceptability. The UK has much to offer in terms of experience and expertise and its membership will enable

UK companies and institutions to participate in research work which may influence commercial opportunities in the future.

(DTI, 2001)

The door to new nuclear was kept open in a report in February 2002 by the Performance and Innovation Unit (PIU), based in the Cabinet Office. This report was produced within the Number 10 team (office of the Prime Minister), independently of the nuclear sceptics in the Trade and Industry and Environment departments. With a foreword by Tony Blair, it showed an important change of tone. It started with a clear statement that energy would be more of a problem in future:

> Trends in energy markets have been comparatively benign over the past 10-15 years: the UK has been self-sufficient in energy; commercial decisions have resulted in changes in the fuel mix that have reduced UK emissions of greenhouse gases; and trends in world markets and domestic liberalisation have reduced most fuel prices.
>
> The future context for energy policy will be different. The UK will be increasingly dependent on imported oil and gas. The Californian crisis has highlighted the importance of putting in place the right incentives for investment in energy infrastructure. And the UK is likely to face increasingly demanding greenhouse gas reduction targets as a result of international action, which will not be achieved through commercial decisions alone.
>
> (PIU, 2002)

The "Californian crisis" refers to a botched deregulation of electricity supply in the state, which was exploited by energy company Enron, leading in 2000 and 2001 to a shortage of power in the peak summer months when air conditioning demand is high, to huge increases in wholesale prices and to the demise of a major utility company. Enron was later found guilty of fraudulent manipulation of the market and went bankrupt in December 2001.

The PIU report signalled the two major energy problems for the UK in future: national security and achieving cuts in greenhouse gases. In discussing this, the report left the door open to nuclear:

> The immediate priorities of energy policy are likely to be most cost-effectively served by promoting energy efficiency and expanding the role of renewables. However, the options of new investment in nuclear power and in clean coal (through carbon sequestration) need to be kept open, and practical measures taken to do this.
>
> (PIU, 2002: "Executive summary". 6)

The official and influential statement of energy policy followed a year later in the DTI's energy white paper (*Our Energy Future*) of February 2003. The paper was unenthusiastic but didn't actually block new nuclear from a future role:

> Nuclear power is currently an important source of carbon-free electricity. However, its current economics make it an unattractive option for new, carbon-free generating capacity and there are also important issues of nuclear waste to be resolved. These issues include our legacy waste and continued waste arising from other sources. This white paper does not contain specific proposals for building new nuclear power stations. However we do not rule out the possibility that at some point in the future new nuclear build might be necessary if we are to meet our carbon targets. Before any decision to proceed with the building of new nuclear power stations, there will need to be the fullest public consultation and the publication of a further white paper setting out our proposals.
>
> (DTI 2003: para. 1.24)

Behind the scenes there was a split between the anti-nuclear Margaret Beckett and Patricia Hewitt, and a growing group of politicians and advisers close to the Prime Minister who thought nuclear needed more support. Within the cabinet in 2003-05, in the lead-up to Labour's third election victory, Gordon Brown (Chancellor of the Exchequer), Alistair Darling (Secretary of State for Transport from 2003) and Alan Milburn (Chancellor of the Duchy of Lancaster and Minister for the Cabinet Office from 2004) were all pro-nuclear, or more accurately, not anti-nuclear. Peter Mandelson, who was in 2004-08 the European Commissioner for Trade, had political influence with the Prime Minister and was also pragmatic on nuclear.

These politicians were all ready to see a role for nuclear power in future energy policy if it was economic. Nuclear was not to be given any special favour but neither was it to be ruled out a priori, either on outdated "green" arguments or because of the regrettable, but largely irrelevant problems of the old nuclear power station designs that had contributed to BE's financial problems. Blair himself was regarded by those around him as pro-nuclear from the start but cautious about saying anything openly on a subject that was controversial with the public ahead of the general election to be held in 2005, when Labour would go for a historic third term. As one adviser put it, from 2004 Labour was leaving the nuclear door open for the private sector to walk through, as it later decided to do.

The reasons for the pragmatic view of nuclear were already in place but were growing in importance. First, North Sea gas was beginning to run out more quickly than expected. This meant the end of the UK's long-running self-sufficiency in primary energy (coal originally, then oil and gas). Second, the closure of coal stations on environmental grounds (at this stage mainly because of regional pollution rather than global climate change) meant new capacity would be needed and the private sector would naturally choose gas. Gas meant lower emissions per unit of power than coal, but the gas would be imported, hurting the balance of payments and energy security. Third, the older, troubled nuclear power stations were closing. Most of the first-generation Magnox stations had closed by 2001 – the last was scheduled to close at the end of 2015. The more modern advanced gas-cooled reactors (AGRs) were approaching the end of their design lives but had some prospect of life extensions, in part because their poor early performance meant their components were less "worn out" than expected. But it was only a matter of time before they were closed too. If the old stations were replaced with gas power stations, they would raise carbon emissions and worsen the UK's energy import dependence. The nuclear stations did have to import their fuel, but uranium was widely available from a number of friendly countries and faced no likely global shortage any time soon.

Labour plans for a third term in government

No Labour government had ever been elected for a third term. Blair now had a real chance of this historic achievement. Aware of how little actually gets done in a government's life, Blair wanted to make sure that the third term was not wasted. He drew on a team in the Cabinet Office that provided the sort of strategic, evidence-based and fearless analysis Blair liked. The crux of this team was John Birt, a long-term associate of Blair's who between 1992 and 2000, as Director-General of the BBC, had controversially reshaped the organization using business school techniques and concepts, which were alien to the BBC. Knighted in 1999, Birt joined the House of Lords as a cross-bencher (not aligned to any party) in 2000.

Blair admired Birt's independent mind and analysis and initially appointed him as a special adviser on the criminal justice system. Birt was the first to say that this was not a subject he knew anything about, but Blair wanted fresh thinking and duly got it. One result of what in business would be called a "deep dive" into the roots of offending was the creation of the Serious Organised Crime Agency (SOCA).

After the 2001 general election, Blair gave Birt an unpaid role as a strategy adviser to the Prime Minister. He produced reports on a number of difficult long-term problems, including drugs, transport, health, education and the London economy. Operating outside the normal civil service framework, Birt's independence inevitably created resistance among some cabinet members and senior civil servants. (Deputy Prime Minister John Prescott once quipped that the unpaid Birt was "worth every penny".) But Birt formed a good working relationship with the Cabinet Secretary, Andrew Turnbull, the single most important non-elected person in the Blair government.

Birt, Turnbull and the head of the No. 10 Policy Unit, Geoff Mulgan, worked together in the three years running up to the general election, which was eventually held in May 2005. Blair asked them to look at a number of subjects that were difficult, controversial and which needed fresh thinking. The goal was to provide Blair with the best chance of achieving something lasting in his third term, one that was expected to be less than five years because at some point Gordon Brown would have to be offered a chance to be Prime Minister.

These thorny subjects tended to be avoided by governments facing day-to-day challenges. One was the structure of government, and the way that the Treasury in particular acted as an independent source of power, especially under a forceful figure such as Gordon Brown. Another was the old chestnut, British productivity, a subject over which hands had been wrung since the late 19th century and something of a totem for would-be modernizing Labour governments. The third was the combination of energy and environmental policy. Although bits of policy had been in place in the previous few years – the renewables obligation and rhetorical support for decisive action on climate change – Blair and the strategy team believed that not enough had been done, and a serious rethink was needed to safeguard energy security and achieve lower carbon emissions. Less prominent but still relevant was concern about how any new policy would affect energy costs for lower-income families, what later became known as affordability.

Anthony Seldon, in the second volume of his biography of Tony Blair, describes how, at a meeting of the key strategy team at the Prime Minister's official country residence, Chequers, Blair asked for a number of five-year plans (Seldon, 2008: Chapter 11). The main ones were on health, education, and law and order. But the team also prepared a report on energy and the environment. This was a more secret piece of work, involving only Birt, Turnbull and a small team from the Policy Unit plus Paul Britton, the Head of the Economic and Domestic Affairs Secretariat in the Cabinet Office. Other civil servants were not involved or even told, though it became generally known that the work was under way and might lead to sensitive, controversial policy decisions.

The secret report revealed, through Birt's relentless focus on data, facts and analysis, that the country faced a serious future risk of a shortfall in power generation, and was also unlikely to hit its renewables targets. Coal and old nuclear were closing. New gas would threaten the greenhouse gas emissions policy. And renewables were nowhere near filling the gap.

The only realistic solution, the team concluded, was to put new nuclear back on the agenda. Nuclear could provide very large amounts of carbon-free energy, replacing the old baseload (continuous generation) coal capacity and leaving space for renewables to grow, while not relying on their intermittent supply to keep the lights on. The only risk in this was the cost – new nuclear looked expensive. But fossil fuel prices were rising and expected to rise further, which made the long-term cost of nuclear seem reasonable, particularly when the benefit of avoided CO_2 emissions was added in.

However, the team still considered nuclear a very controversial subject. It was not clear how many cabinet members were opposed to nuclear, or at best acquiescent. And the general public were regarded as likely to revolt. Blair, ever cautious despite encouraging opinion polls going into the last year ahead of the election, said nothing public about nuclear power.

Blair had apparently considered openly telling the public that he was in favour of nuclear energy as early as 2002, but his aides had warned him off. Two people who helped him rediscover his support for nuclear, based on facts and analysis, were Sir David King, a Cambridge University engineer who was the government's Chief Scientific Adviser in 2000-08, and Sue Ion, Chief Technology Director at BNFL in 1992-2006, and one of the UK's leading nuclear scientists.

King found Blair to be one of those "non-specialists who were intrigued by science" (Seldon, 2008: chapter 17). Blair found King a credible and persuasive advocate for nuclear power and asked him for additional reading material on the case for nuclear being safe and reliable.

Sue Ion was a member of the UK Panel on Science and Technology from 2004 to 2011. The panel provides cross-departmental advice to the Prime Minister and is chaired by the Chief Scientific Adviser. Ion was therefore meeting Blair every six months and was able to press the case for nuclear research (she controlled one of the few remaining nuclear research budgets, that at BNFL) and for new nuclear power investments. A BBC Radio 4 interview in 2013 described her persuading Blair in 2006 to look again at nuclear, but in truth she had already helped Blair to reach this decision, before the general election in 2005 (BBC, 2013). One of the panel's reports looked at the UK energy outlook 15 and 20 years out. It was the sort of long-term thinking that Blair respected. The report made it clear that most of the older nuclear reactors would be closed by then and new nuclear would be essential just to replace them, unless carbon emissions and energy security were both to worsen.

The Labour manifesto in 2005 read:

> Our wider energy policy has created a framework that places the challenge of climate change – as well as the need to achieve security of supply – at the heart of our energy policy. We have a major programme to promote renewable energy, as part of a strategy of having a mix of energy sources from nuclear power stations to clean coal to micro-generators.
>
> (Labour party, 2005: 22)

The Conservative manifesto didn't contain the word nuclear:

> A Conservative Government will guarantee the security and sustainability of Britain's energy supplies. We will do this by supporting the development of a broad range of renewable energy sources. We also recognise that energy efficiency must play an increasingly important role in our energy policy.
>
> (Conservative party, 2005: 23)

Blair won his third term. After officially meeting the Queen at Buckingham Palace on the morning of Friday 6 May (his birthday), a tired Blair, demoralized by the reduced majority and continual attacks on his Iraq invasion decision, returned to No. 10 Downing Street. His first appointment was with Birt and Turnbull (Seldon, 2008: Chapter 13). Over the next few days it became clear that Blair's dependence on Gordon Brown, who opposed much of the more radical strategic thinking that Birt and Turnbull had been working on, meant a lot had to be scrapped. An attempt to change the name of the Department of Trade and Industry to include the words "productivity" and "energy" foundered on concerns that the name could too easily be abbreviated to PENIS. But one area that survived was the semi-secret energy plan.

The UK sells its nuclear reactor business (2005)

Around the same time that the Prime Minister was beginning to conclude that new nuclear power was essential to the UK's energy future, another part of the government was trying to dispose of one of only three or four manufacturers of new nuclear reactors, a company which belonged to the British taxpayer.

Westinghouse Electric was a historical American industrial company, one of the pioneers of alternating current (the system now universally adopted for power generation and transmission but locked in contention with the competing direct current system used by George Westinghouse's great rival, Thomas Edison). Westinghouse won the "war of the currents" in the late 19th century and even after George's death in 1914, went on to become a major electrical engineering company, continuing to compete with Edison's General Electric (GE).

Decades later, Westinghouse became the main supplier of what became the world industry-standard reactor, the pressurized-water reactor (PWR). More than 270 PWRs have been built across the world, the majority being either based on or derived from the Westinghouse design. The UK has a single Westinghouse PWR at Sizewell in Suffolk. France licensed Westinghouse designs and then developed its own advanced version, the European pressurized-water reactor (EPR). China in turn licensed the French PWR design and developed its own more advanced model, the CPR1000.

When US nuclear construction ended in the 1980s, Westinghouse's nuclear business shrank, leaving only the waste and fuel activities. In a remarkable corporate strategic transformation, Westinghouse in 1995 bought one of the three main US broadcasting companies, CBS. It then changed the company name to CBS and disposed of its non-media activities, selling its nuclear business to the British government-owned BNFL in 1998.

BNFL was the company responsible for the nuclear wilderness at Sellafield, plus the older Magnox nuclear reactors. The Magnoxes were so close to the end of their lives that they had a small or negative economic value once their decommissioning liabilities were taken into account. They couldn't be privatized so they had been given to BNFL in the mid-1990s. The government still hoped to somehow privatize BNFL but it faced a difficult challenge. Quite apart from the controversial nature of its business, BNFL was a company with some, rather tentative, cash flow growth derived from its expertise in cleaning up nuclear waste and, even more dubiously, from using its plutonium stockpile to manufacture MOX fuel. Lacking any reliable growth potential made it unappealing to investors. So the British government in 1998 authorized BNFL to buy Westinghouse's nuclear operations, including the AP600 "advanced passive" reactor design newly approved by the US Nuclear Regulatory Commission. This gave BNFL a stake in the prospective global resurgence in nuclear construction, which was based on new reactor designs such as the AP600 and its larger version, the AP1000. These reactors were designed to fail safely; even with a total loss of coolant and power the reactors would shut down safely with no risk of either meltdown or radiation emission.

In 2000 BNFL also bought the nuclear activities of the Swiss company ABB for £300m and acquired a minority stake in a South African company that had developed a pebble bed modular reactor. (Through the years of apartheid South Africa had invested heavily in potential sources of independent energy to frustrate the international oil embargo.)

With these stakes in a growing nuclear future, albeit one that the UK in the early 2000s seemed unlikely to take part in, the company seemed to have a chance of being privatized. But the MOX scandal in 1999 (see above) and continued financial difficulties linked to both the US and the company's attempts to revive the idea of a full-scale MOX plant, the government gave up on privatization. The clean-up parts of BNFL would have to continue into the very long-term future, but it appeared to make no sense to continue to own Westinghouse with its promising new reactor technology.

So in 2005, just as the Blair government was behind the scenes preparing to affirm the need for new nuclear build in the UK, the UK-owned BNFL sold Westinghouse – one of the most likely contenders for such investment. BNFL reportedly begged the government to delay the sale by a year, amid growing expectations for a future nuclear renaissance worldwide. The sale attracted much more interest than expected and Westinghouse was bought by the Japanese company, Toshiba, for $5.4bn. This put Toshiba back in the running against its rival Hitachi, which had developed an advanced version of the BWR pioneered by Westinghouse's old rival, GE.

In 2007 Toshiba submitted its AP1000 reactor plans to the UK Office for Nuclear Regulation (ONR), to start the process of generic design assessment. Toshiba described it as "the safest and most economical nuclear power plant available in the worldwide commercial marketplace", though that is naturally contested.

Did the UK miss an opportunity to hold a stake in the nuclear reactor business? Perhaps; but most of the AP1000 supply chain is overseas. And there were good policy reasons for the UK government to get out of the reactor business, which it had entered only for short-term, tactical reasons. It would have been very difficult for the AP1000 design to compete in the UK if it had been owned by the government, since everybody would have suspected bias in any decisions taken, even if the end-customer had been a private utility. It would also have made it harder for Westinghouse to sell abroad. The AP1000 is now under construction at two sites in China and is a serious contender for a future site at Sellafield, close to the BNFL complex. This might not have happened if the company had been constrained by government ownership.

Conclusion

In the middle of 2005 the third Labour government was about to start the process of building a pro-nuclear energy policy. The UK had just sold its only power station nuclear reactor business, Westinghouse. And 2005-07 would see a near-consensus in the UK that new nuclear stations were not just acceptable but unavoidable. What was at first a separate process – the passing of the landmark Climate Change Act – would have the twin distinction of uniting nearly all shades of political, environmental and business opinion across the country, while being largely ignored by the mass of the population.

REFERENCES

BBC Radio 4 (2013). *The Life Scientific*. http://www.bbc.co.uk/programmes/b01qw9hj
The Conservative party (2005). *Manifesto*. http://www.politicsresources.net/area/uk/ge05/man/manifesto-uk-2005.pdf
DTI (2001). *Energy Report*.
DTI (2003). *Energy White Paper: Our Energy Future*. Cm5761. February. http://webarchive.nationalarchives.gov.uk/+/http:/www.berr.gov.uk/files/file10719.pdf.

Performance and Innovation Unit (PIU) (2002). *The Energy Review.* February. http://www.gci.org.uk/Documents/TheEnergyReview.pdf

The Labour party (2005). *Manifesto.* http://www.politicsresources.net/area/uk/ge05/man/lab/manifesto.pdf

Seldon, A. with Snowdon, P. and Collings, D. (2008). *Blair Unbound.* London: Pocket Books.

The Times (2005). "The nuclear cabinet" 23 November.

5 New nuclear build becomes official policy
(2005-08)

Blair becomes fully pro-nuclear

Under the influence of David King, Sue Ion and the work done by his strategy team, Tony Blair had come to the conclusion, even before his third general election victory of May 2005, that new nuclear investment was inevitable. But he remained cautious in public.

In 2005 it was the UK's turn to host the increasingly important annual G8 leaders' meeting. This summit had grown from being a largely technical affair to the most important meeting for the leaders of the world's leading advanced economies, with Russia included out of courtesy, it not being an "advanced" economy in this context. Blair, as host, was keen to make the event at Gleneagles a landmark by achieving a number of popular goals, such as forgiving poor country debt and increasing aid to Africa. He also wanted some real progress on climate change.

He didn't get everything he wanted and the meeting was interrupted by the 7 July 2005 terrorist bombings in London. The G8 action plan on climate change issued on 8 July, following the summit, barely mentioned nuclear and gave the impression that it was not central to UK policy:

> We take note of the efforts of those G8 members who will continue to use nuclear energy, to develop more advanced technologies that would be safer, more reliable and more resistant to diversion and proliferation.
>
> (G8 (2005) para. 12)

In his speech to the Labour party conference on 27 September 2005, Blair laid out his agenda for the new government. After education, expanded child care, health and pensions, he came to energy and for the first time opened the door to an expanded role for nuclear power:

> Global warming is too serious for the world any longer to ignore its danger or split into opposing factions on it. And for how much longer can countries like ours allow the security of our energy supply to be dependent on some of the most unstable parts of the world? For both reasons the G8 Agreement must be made to work so we develop together the technology that allows prosperous nations to adapt and emerging ones to grow sustainably; and that means an assessment of all options, including civil nuclear power.
>
> (Blair, 2005)

The Royal Academy of Engineering had been running a series of policy seminars on energy matters. In September 2005 it addressed nuclear power. A later report summarized the seminar as finding that "the often quoted concerns over safety, economics and waste management are over stated" and "Western nuclear reactors are now becoming available that are simplified in design and provide enhanced protection against safety related failures." The report argued that, for the private sector to be willing to deliver new nuclear power stations, the government needed to provide long-term assurances on carbon pricing and nuclear regulation and to improve the planning process (Royal Academy of Engineering, 2006).

In October the Engineering Employers Federation (EEF), which represented many manufacturing companies that were major users of electricity, published a report, *Sustainable Energy – A long-term strategy for the UK*. The report recommended combining renewable and new nuclear power to achieve competitive low-carbon power supplies.

The same month, at its annual conference, the Confederation of British Industry (CBI), which represents a wide group of businesses, urged the government to take a decision within a year to allow new nuclear build.

On 21 October Sir David King told the *Guardian* that Britain would need to revive its nuclear power industry, for both economic and environmental reasons (*Guardian*, 2005). He pointed out that the closure of old nuclear stations meant that some 20% of the UK's current electricity production (and most of its low carbon power) would disappear. It would be very difficult to replace it without new nuclear build, unless emissions were to rise.

On 20 November he went further and told the BBC's AM programme that it was time "to give the green light to the private sector and the utilities and give them nuclear as an option" (BBC News, 2005). He described the UK's nuclear share of electricity as 21%, down from 24% a few years before and on track to be only 4% by 2010 as older stations closed (although this would turn out to be too pessimistic). A major source of low-carbon power was in decline. Giving the private sector the nuclear "option" implied that nuclear would be voluntarily delivered by the private sector, without subsidy. This was the form of words that continued into the nuclear white paper of 2008, although King had made it clear in other interviews that the economics of new nuclear depended on an appropriate price being put on carbon emissions.

Sir David's comments were also reported in the *Western Morning News*, based in Bristol, where speculation was building about the revival of the Hinkley Point C (HPC) new nuclear project, down the coast in Somerset.

Also in November, Dr John Loughead, Executive Director of the recently established UK Energy Research Centre, a body of scientists funded by the various government research funding councils, authored a report that drew on work by some 150 scientists and energy experts. It concluded that the UK's looming energy gap would need to be filled in part by new nuclear power:

> Nuclear fission energy is a proven and reliable technology that will inevitably have a key role in a future clean energy mix.
>
> (Institute of Physics, 2005)

The flow of support for new nuclear had the appearance of a coordinated campaign but was really the result of a growing expectation that the government was serious about new nuclear power and would soon make an important change in policy. In particular, it was widely believed that Blair was himself in favour of nuclear power.

These expectations came to fruition on 29 November 2005, when Blair announced a new energy review led by the Department of Trade and Industry (DTI) that would include the option of new nuclear power stations. Finally going public on an issue that he feared would be highly controversial, Blair's concerns seemed at first correct. His speech to members of the CBI in London had to be delayed when two Greenpeace activists climbed inside the roof of the Design Centre in Islington, the location of the meeting, and unfurled banners that said "Nuclear: Wrong Answer". Press coverage of the announcement reported opposition from some 50 Labour and Liberal Democrat MPs, including the former Environment Minister, Michael Meacher. The Friends of the Earth (FoE) Director, Tony Juniper, commented more moderately that the UK could meet its emissions and energy security goals using "clean, safe alternatives" (*Telegraph*, 2005).

The review, under Minister of State for Energy, Malcolm Wicks, published a consultation document in January 2006, which gave three reasons to revisit energy policy:

- Continued evidence of the damage from climate change;
- Energy security: the UK had become a net importer of gas sooner than expected and would soon be a net importer of oil;
- Energy prices had risen sharply and threatened to reverse the fall in the number of people in fuel poverty. The number had dropped from 6.5m in 1996 to 2m in 2003.

(DTI, 2006a)

The consultation document was short, offered very little detail and required comments by 14 April, less than three months later.

As if to reinforce the point about energy security, on New Year's Day 2006 the Russian gas giant Gazprom started to reduce the pressure in its gas pipeline supply into Ukraine, in one of a continuing sequence of disputes on payment. Although the dispute was only with Ukraine itself, the country was the main route to the rest of Europe for Russian gas, and soon other European countries were affected by the Russian action.

The UK did not take any gas directly from Russia (nor did it in 2015), relying instead on its own North Sea resources, on the Norwegian and Dutch North Sea sectors, and on liquefied natural gas (LNG) from many other countries. However, the UK was part of the European gas market, to which it was connected via pipelines. So disruption to supplies of gas in the rest of Europe was still a concern for the UK.

The consultation document included a goal of putting the country "on a path to cut the UK's CO_2 emissions by about 2050, with real progress by 2020". This was the first time that this long-term target had been expressed as in any definite way, and reflected the growing momentum behind codifying government strategy in a specific, quantitative policy.

The consultation document also noted that, on current trends, emissions would be barely lower by 2020 and that a lot of generating capacity was scheduled to retire in the next few years, requiring new investment. While stating that "no single measure can deliver our goals" it highlighted a review of nuclear's contribution as the main new topic. The 2003 energy white paper from Margaret Beckett had in effect dismissed nuclear on the grounds of economic cost and the unresolved problems of nuclear waste. But with rising fossil fuel prices the 2006 review wanted to reconsider these arguments. It invited answers and comments on five questions, one of which was whether there were "particular considerations" that applied to nuclear as the government re-examined the question of new build.

Although this was a consultation document and part of a wider review, the process was widely seen as signalling a change in government policy, since the answers to the five questions were likely to be in line with the many pro-nuclear documents published in the previous year.

That was certainly the view from the electricity industry. In April 2006 Mike Parker, Group Chief Executive for BNFL, publicly stated that the French power giant, EDF and the two major German utilities, E.ON and RWE had expressed interest in building new nuclear power stations (*Nucleonics Week*, 2006).

The momentum towards nuclear increased when on 12 May David Miliband became the new Environment Secretary, after Blair changed his mind about appointing the relatively inexperienced Miliband as Foreign Secretary and appointed a surprised Margaret Beckett instead. Whatever the political and personal reasons for the change, the replacement of anti-nuclear Beckett by pro-nuclear Miliband was reported by the BBC as "widely seen as clearing one obstacle to building more nuclear plants" (BBC News, 2006).

The energy review was parented by the DTI, though. In July 2006 the review's conclusion was published as *The Energy Challenge* (DTI, 2006b). In his foreword, Tony Blair summarized the nation's need to invest in renewable energy sources and to improve energy efficiency. But, he said, this would not be enough. The country would also have to find new sources of energy abroad, to find ways to burn coal and gas cleanly and to invest in new nuclear power capacity.

The Energy Challenge made it clear that the burden of construction, financing, waste disposal and decommissioning lay with the private sector. Quoting the interim report of the Committee on Radioactive Waste Management, published in April 2006, *The Energy Challenge* said that deep geological disposal of waste was feasible, although the committee's report was a long way from providing a settled and workable policy. The energy review's final report, published at the end of July, after *The Energy Challenge*, reaffirmed this general conclusion but emphasized the "uncertainties surrounding disposal" and recommended more research (CoRWM, 2006). The question of waste disposal remained unresolved years later, and yet has not stopped the development of new nuclear power stations.

The review promised a white paper by the end of 2006 and further exploration of national energy strategy and regulation issues, but recommended no public inquiry.

Those who had waited decades for the return of nuclear power were jubilant. It seemed that at last the economic and political stars were aligned in favour of new nuclear investment in the UK. Private sector utilities were talking openly of their interest in building, without subsidy (though with some form of carbon pricing), new stations that were safer, more efficient and that produced a lower volume of waste than the previous reactors. These stations would provide economic and carbon-free energy, without the UK depending on gas imported from potentially hostile or unreliable suppliers.

Blair's caution over nuclear power had been due to concerns about public reaction, and it was not yet clear what shape this would take. And while the review had included a consultation exercise, there were concerns about how credible this exercise was. Just before the energy review's full report was published, the House of Commons Trade and Industry Select Committee published its own report, called *New Nuclear? Examining the issues*, which expressed strong concern about the consultation process for "one of the most important issues [the government] has faced in its time in office" (House of Commons, 2006).

New Nuclear? concluded that a government policy to enable the construction of new nuclear power stations would only be credible if it was based on four elements:

i)   A broad national consensus on the role of nuclear power, which had both cross-party political support and wider public backing;
ii) A carbon-pricing framework that provided long-term incentives for investment in all low-carbon technologies;
iii) A long-term storage solution in place for the UK's existing radioactive waste legacy;
iv) A review of the planning and licensing system to reduce the lead time for construction.

New Nuclear? was not critical of the policy of new nuclear itself, indeed it appeared broadly supportive. Rather, it pointed to a number of practical matters that would need to be addressed if new nuclear were ever to happen. These included: sites for new stations; new and in some cases unproven reactor types; the considerable financing challenges; and hurdles in the engineering supply chain. All of these concerns turned out to be accurate.

The committee's criticisms of the weakness of the process would also turn out to be prescient. Conservative MP Peter Luff, Chairman of the committee, told the *Guardian* in July that the energy review "risks being seen as little more than a rubber-stamping exercise for a decision the Prime Minister took some time ago" (*Guardian*, 2006).

Peter Luff was not the only one to think this. On 6 October 2006 Greenpeace, still relentlessly hostile to nuclear power, mounted a legal challenge to the energy review. It claimed that the government had failed to live up to its earlier commitment in paragraph 4.68 of the 2003 white paper that "Before any decision to proceed with the building of new nuclear power stations, there would need to be the fullest public consultation and the publication of a white paper". In particular, Greenpeace argued, the energy review had barely consulted the nation on the questions of management of radioactive waste, the siting of new stations and their cost (Greenpeace, 2006).

The government was not surprised by Greenpeace's legal challenge but didn't think it was likely to lead anywhere, though some civil servants privately thought that Greenpeace had a point. Meanwhile the broader momentum behind climate change policy increased markedly with the publication on 30 October 2006 of the *Review on the Economics of Climate Change* by Professor Nicholas Stern of the London School of Economics. This (commonly known as the Stern review) had been commissioned by Gordon Brown, who was competing with Blair throughout the early years of the third Labour government. But climate change was one issue on which the rivals Blair and Brown broadly agreed.

The Stern review was a detailed, serious and credible examination of the economic costs and benefits of climate change, and policy to halt or reduce it. It provided strong evidence that government action was required.

Among its conclusions were: "Using the results from formal economic models, the Review estimates that if we don't act, the overall [global] costs and risks of climate change will be equivalent to losing at least 5% of global GDP each year, now and forever. If a wider range of risks and impacts is taken into account, the estimates of damage could rise to 20% of GDP or more." It added: "In contrast, the costs of action – reducing greenhouse gas emissions to avoid the worst impacts of climate change – can be limited to around 1% of global GDP each year" (Stern, 2007). This estimate was based on the assumption that concentrations of greenhouse gases in the atmosphere would be stabilized at between 450 and 500 parts per million of carbon dioxide equivalent.

Stern's review was criticized by some economists, who focused on the critical but contentious question of how to convert future costs and benefits to values in today's money, otherwise known as discounting. Aside from this very technical matter, the review was widely applauded and set a new benchmark for the intelligent discussion of climate change policy. In particular, it provided a justification for taking action, as the current costs of such action would be far less than the future costs of inaction: a standard economic cost/benefit calculation.

Greenpeace blocks the nuclear review

In early 2007, as the climate change bill gathered momentum (see below), the progress of nuclear came to a sudden halt. On 15 February, the High Court ruled in favour of Greenpeace's legal challenge, finding that the government had failed to consult properly ahead of the new pro-nuclear policy published in the July 2006 energy review.

The judgement was damning. Mr Justice Sullivan said "something has gone clearly and radically wrong". He said that the consultation process had been "seriously flawed" and "procedurally unfair". The consultation document had given every appearance of being simply an "issues paper". It had contained no actual proposals and the information given to consultees had been "wholly insufficient for them to make an intelligent response". Even worse, the judge said the information given on radioactive waste had been "not merely inadequate but also misleading".

He concluded that fairness required that the consultees should be given a proper opportunity to respond to the substantial amount of new material before any decision was taken. In sum, "There could be no proper consultation, let alone the fullest consultation, if the substance of these two issues was not consulted on before a decision was made" (Tromans (2010) Chapter 5).

The ruling came as a shock to the government. Blair defiantly told the House of Commons that, while the ruling would change the consultation process, "this won't affect the policy at all". He restated that new nuclear power stations were necessary to hit climate change targets and to provide energy security (BBC News, 2007).

The ruling, while highly critical of the government, was none-theless only about process and procedure. The government was free to make policy as it wished, in line with its democratic mandate. But it could not promise to consult and then fail properly to do so. There was therefore every reason to believe that the policy of building new nuclear stations could go ahead but the government would have to conduct a more convincing consultation. In hindsight, that was a blessing for the government. It meant that when new nuclear became a real possibility, the government could not be accused of pushing the policy through covertly or without telling the public what was going on. Greenpeace, it would turn out, had forced the government to make their major new policy much more robust to future public criticism.

A real consultation (2007)

On 23 May 2007 the government published both a new energy white paper and a new consultation report. In a foreword to the white paper, *Meeting the Energy Challenge*, the Secretary of State for Trade and Industry, Alistair Darling wrote: "we are publishing a consultation document on nuclear power so that we can take a decision before the end of the year on whether it is in the public interest for companies to have this option available when making their investment decisions."

The government now stated three strategy goals:

i)   Saving energy;
ii) Developing cleaner energy supplies;
iii) Securing reliable energy supplies at prices set in competitive markets.
(DTI, 2007a)

Note that the reference to competitive markets meant the government continued to see energy delivery as primarily the private sector's responsibility, guided by appropriate market interventions such as a price on carbon.

The consultation report was called *The Future of Nuclear Power*. At more than 200 pages it was a very different document from the 2006 consultation, in scale, scope and tone.

It started by asking how nuclear would fit in with the new energy strategy. About one-third of all UK power generation capacity was due to close in the next two decades, and half of this (10 GW) was nuclear. If the nuclear capacity were to be replaced with generation from fossil fuels instead, it would use up 30-60% of all the carbon savings projected to arise from the various measures proposed in the energy white paper. It would also mean higher gas demand, increasing the UK's need to import energy from abroad. So *not* replacing old, closing nuclear stations with new ones would undermine both the cleaner energy goal and the energy security goal.

The report went on to say that nuclear involved long lead times. The government "conservatively" assumed a lead time of eight years pre-construction and five years for construction of the first new plant, dropping to five and five, respectively, for later stations (footnote 10). Rather than conservative, many commentators regarded those times as realistic or even slightly optimistic; but they were based on what the private sector was then telling the government. The new Department of Business, Enterprise and Regulatory Reform (BERR), carved out of the abolished DTI in June 2007, separately published a cost/benefit analysis of the case for new nuclear stations, which assumed six years for construction (BERR (2007a) p. 1).

So new nuclear was unlikely to come on stream much before 2020. "New nuclear power stations could make an important contribution . . . to meeting our needs for low carbon electricity generation and energy security in this period and beyond to 2050. Because of the lead-times, without clarity now we will foreclose the opportunity for nuclear power."

The consultation report described explicitly what had changed since the 2003 white paper, which had marked nuclear as uneconomic and bringing "unresolved issues of nuclear waste". Since then, the new consultation document said, "there has been significant progress in tackling the legacy waste issue". It claimed that there were now technical solutions for waste disposal that scientific consensus and experience from abroad suggested could accommodate all types of wastes from existing and new nuclear power stations.

The government had also created an implementing body, the Nuclear Decommissioning Authority (NDA), which had expertise in this area, and was reconstituting the Committee on Radioactive Waste Management (CoRWM) to provide continued independent scrutiny and advice. It promised a framework for implementing long-term waste disposal in a geological repository to be consulted on in the coming months.

Again, this seemed to overstate progress. There was indeed wide-spread agreement internationally that deep geological storage was the best way of dealing with long-term nuclear waste. But identifying a site with a willing local population was (and remains) some way off.

With respect to new nuclear stations, the consultation paper invited comments on the "ethical, intergenerational and public acceptability issues" arising from allowing the building of new stations that would generate additional nuclear waste. It also promised to ensure that the private operators of any new stations were fully responsible for the costs of decommissioning and waste storage. This was the policy in operation for British Energy (BE) when it was privatized but the new arrangements would be more robust.

Regarding the economic argument, the consultation stated that "The high-level economic analysis of nuclear power, prepared for the energy review, concluded that under likely scenarios for gas and carbon prices and taking prudent estimates of nuclear costs, nuclear power would offer general economic benefit to the UK in terms of reduced carbon emissions and security of supply benefits." In effect this amounted to saying that rising fossil fuel prices, together with a mechanism for adding in the damage done by carbon emissions, would make nuclear economic compared with the gas power stations, based on private sector estimates of nuclear construction costs.

Unlike in 2003, there were now energy companies expressing a strong interest in investing in new nuclear power stations. "They assess that new nuclear power stations could be an economically attractive low-carbon investment, which could help diversify their generation portfolios. Their renewed interest reflects assessments that with carbon being priced to reflect its impacts and gas prices likely to be higher than previously expected, the economics of new nuclear power stations are becoming more favourable."

The government's confidence in the prospect of new nuclear was substantially underwritten by private sector confidence in the economic benefit of new nuclear. That confidence in turn depended on an effective mechanism for carbon pricing and on the government making changes to the planning and regulatory regime that would make planning, building and running new stations quicker and more predictable than in the past.

The government then offered its preliminary view that policy should open up the full range of low-carbon options, including nuclear. It cited the finding in the recently published fourth report of the Intergovernmental Panel on Climate Change (IPCC) that nuclear power could have a role to play alongside other low-carbon energy sources in reducing carbon emissions:

> Our preliminary view is that preventing energy companies from investing in new nuclear power stations would increase the risk of not achieving our long-term climate change and energy security goals, or achieving them at higher cost.

Note that the view was expressed as allowing private sector companies to make commercially motivated decisions. This is very different from the government telling or pushing an unwilling electricity industry to make investments that would not otherwise happen.

The consultation document then went on to flesh out the government's arguments on each point, together with reassurances on non-proliferation and transportation of waste, in each case asking the public whether it agreed or disagreed, and if so, why? It also covered the supply of nuclear fuel, supply chain matters, the government's view that in future spent fuel should not be reprocessed (unlike in the past), planning changes, questions of siting, and asked whether there was any other relevant matter that the public thought should be considered.

The government had indeed taken full account of the legal ruling. It would be hard to argue that this new consultation document was not comprehensive, detailed and specific. It laid out arguments on all the main matters, expressed the government's view and invited comment. The consultation opened on 23 May 2007 and closed on 10 October 2007 (compared with three months for the previous consultation). There were to be nine "citizen deliberative events" held in September in Belfast, Cardiff, Edinburgh, Exeter, Leicester, Liverpool, London, Newcastle and Norwich.

Looking back, the public acceptance of and political consensus on new nuclear power owe much to this consultation. By inviting comments on every aspect of its nuclear proposals, the government undermined any later claim that it had failed to involve the nation in this new policy.

The private sector plans for new nuclear stations

BE had announced in February 2007 that it was inviting potential partners to submit proposals to build new nuclear power plants in anticipation that the government would give the firm the go-ahead.

The value of its shares now depended not just on the value of the existing, frail assets but on the potential for new investment. BE's sites had additional value as the most likely locations for new investment, as construction could take place without raising new planning problems.

Despite a run of bad news about the older stations, in December 2007 BE received permission from the nuclear regulator to extend the lives of two reactors – Hinkley Point and Hunterston – by an extra five years to 2016. This temporarily reduced the forecast fall in UK nuclear output, but there would still be a fall before new nuclear would come online (in 2020 or later).

In November 2007 BE announced it had signed connection agreements with National Grid for four new nuclear stations: Bradwell B, Dungeness C, Hinkley Point C (HPC) and Sizewell C, to start in 2016 (Platts commodity news, 2007). This was a long way from agreeing to actually build anything. Among the many conditions for a new power station, it must have access to the transmission grid run by National Grid, a private but regulated company.

National Grid plans a long way ahead and needs to know of any major new supply proposed, to make sure the grid can sustain the extra load. If the supply is too great for the existing capacity, extra power lines and/or control equipment may need to be added. If there is an imbalance (too much supply in one part of the system), then the grid may wish to alter the sequence in which stations are added.

Despite this, the plans were the first tangible evidence that the electricity supply system was beginning to plan for new nuclear power stations, somewhat prematurely as it turned out. It also showed the expected order in which new stations might be added. All were existing nuclear sites, though Bradwell, in Essex, was the site of a first-generation Magnox station (station "A"), which had closed in 2002. This meant that they probably had a high degree of local public acceptance of nuclear, which provided a lot of well-paid jobs in otherwise rural or semi-urban areas where most of the alternative jobs were in the relatively low-paid tourism sector.

The existing sites also had the essential infrastructure for construction access and of course were already connected to the grid. All were in the south of the country, where the demand was greatest relative to supply. Two were on the sites that had come closest to new construction in the early 1990s, before the government halted all new nuclear construction: namely Hinkley Point in Somerset and Sizewell in Suffolk.

On 26 November the CBI's climate change task force published a report, *Climate Change: Everyone's business*. The task force, made up of British companies between them employing nearly two million people, called for "a much greater sense of urgency" in British climate change policy. Among its proposals, it backed the government's nuclear policy, calling for 12 new nuclear stations to be built by 2030, which would require government decisions to be taken before 2011 (CBI, 2007).

The Institute of Directors (IoD), representing smaller companies in the UK, published its own report in November, which accepted the need for action to combat climate change but worried about the effect of poorly designed policies on the British economy. It argued that good environmental credentials should not be established at the expense of damage to the UK's economic competitiveness. The worst way to reduce carbon emissions, it said, would be to cut jobs in the UK and send emissions abroad, often to locations that are less environmentally sensitive than here in the UK. "It would be foolish to decimate parts of British industry in the pursuit of cuts in emissions that are a fraction of the annual growth of emissions in China" (IoD, 2007). The IoD was one of a very few organizations taking a more sceptical view of the global weight of British efforts at this point.

The nuclear consultation closed on 10 October 2007. When the government published its response in January, it described the consultation process in detail. There were 2,728 responses, of which 1,784 were from individuals. The consultation website received 46,000 unique visitors. Some 949 people attended the 9 deliberative events. A stakeholder workshop in July had involved organizations chosen and invited by an independent convenor to come to London in person, including Greenpeace, SERA campaign group, Friends of the Earth, Environment Agency, Nuclear Industry Association, BE plc, EDF Energy, GMB union, Unite union, Nuclear Decommissioning Authority, CBI, Renewable Energy Association, Cardiff University, UK Sustainable Development Commission, COI communications agency, BERR, Opinion Leader and Shared Practice (the independent evaluator; attending in an observer capacity).

A Citizen's Advisory Board of 12 members of the public met 3 times. A development event was held in August, involving 30 members of the public chosen to be demographically representative. A further 13 public meetings were held in the regions and devolved administrations (Scotland, Wales and Northern Ireland). Additional meetings were held in areas likely to be affected by new nuclear construction. The consultation advertised in print and through Google sponsored-search and sent nearly 4,500 direct mailings to grass roots organizations. The cost of the whole consultation process was £2.4m (BERR, 2007b).

Nobody surely could object that this had been a flawed or inadequate consultation process. There would be no legal challenge to this process, though Greenpeace later made unsuccessful challenges in 2011 and 2014, in relation to the Fukushima nuclear accident and the European Commission's state aid approval statement, respectively.

The UK officially approves new nuclear

Prime Minister Gordon Brown, in the foreword to a new white paper published in January 2008, announced what was never really in doubt: "We have therefore decided that the electricity industry should, from now on be allowed to build and operate new nuclear power stations, subject to meeting the normal planning and regulatory requirements" (BERR, 2008). The post-2006 rhetorical style continued: the private sector was champing at the bit to build new nuclear stations; why should the government stop this when it would contribute to the twin policy goals of low-carbon power and energy security?

John Hutton, the Secretary of State for BERR, acknowledged in the white paper that there remained "widespread concerns" about nuclear power, and that to address them, the government would develop:

i)   A clear strategy and process for medium- and long-term waste management, "with confidence that progress will be made";
ii) New legislation for a funding mechanism to ensure that operators of new nuclear power stations made sufficient and secure financial provision to cover their full costs of decommissioning and waste management;
iii) A further strengthening of the resources of the Nuclear Installations Inspectorate (NII – the nuclear regulator) to enable it to meet a growing workload.

These were actions the government was going to take anyway. The Treasury, having been forced to pick up the bill for BE's nuclear liabilities after the company, in effect, went bankrupt, was not going to let that happen again. The NII needed more people to start the onerous process of checking the new nuclear reactor types that would compete for the first new station. On waste the policy was, well, to have a strategy. That wording turned out to be important: the development of new nuclear stations was allowed, even when there was still no practical solution to the waste problem.

The Climate Change Act of 2008 accidentally boosts nuclear

The Climate Change Act, which commits the UK to cutting total carbon emissions by 80% by 2050, was passed with near-unanimous support from MPs in October 2008. It is a remarkable piece of legislation in its ambition, in the implications it has for government policy and in its attempt to overcome a key feature of the British constitution: namely that a parliament cannot bind its successors.

It started life as a campaign by FoE, following a 2004 landmark speech on climate change by Tony Blair. The "big ask" campaign's goal of legally binding targets on carbon emissions led to a private members' bill in 2005 and had by late 2006 gained cross-party consensus, with strong support from the new Conservative Leader (and later Prime Minister) David Cameron. The Labour government's draft bill in 2007 proposed:

   i) An emissions reduction pathway to 2050, with legally binding medium- and long-term targets in the form of five-year carbon budgets;

   ii) Establishing a new Committee on Climate Change (CCC), which would report publicly to Parliament on progress;

   iii) New powers to allow the government to achieve the targets, including additional emissions trading schemes.

(DEFRA, 2007)

The bill converted the 2003 aspiration into a legal commitment to cut CO_2 emissions by 60% by 2050, from 1990 levels. The overwhelming response from the CBI, from the various parliamentary committees and from environmental groups was positive.

Many bills are gradually diluted in their passage to final Royal Assent. But the climate change bill became stronger. On 23 September 2007 Gordon Brown, who had taken over from Blair as Prime Minister on 27 June 2007, told the Labour party conference that he thought the target should be raised to 80%.

On 7 October 2008 the interim CCC recommended that the bill should indeed be amended to 80% cuts by 2050. Just over a week later, the new Environment and Climate Change Secretary, Ed Miliband, confirmed that this would happen. On 26 November 2008 the bill received Royal Assent to become the Climate Change Act.

It is doubtful whether the UK has published any peacetime legislation to rival the scope of the act, though the British public were distracted by the unfolding global financial crisis and may not have noticed. It is also unclear whether the authors and supporters of the act gave much attention to the consequences: namely just how could such a dramatic decarbonization of the British economy be achieved? Could it be done without a large increase in the share of nuclear power? The CCC was later to conclude that nuclear would be an essential part of the solution.

Conclusion

The policy of building new nuclear stations depended on the private sector wanting to invest, encouraged by government incentives that reflected the benefits of low-carbon power.

The Climate Change Act of November 2008 committed the government to achieving increasingly stringent carbon emission cuts. This later reinforced the policy of building new nuclear stations. But could the government rely on the private sector to do what was needed to achieve these goals? What would be the price of private investment?

REFERENCES

BBC News (2005). "Environment shuffle?" 20 November. http://news.bbc.co.uk/1/hi/programmes/sunday_am/4454174.stm

BBC News (2006). "Miliband 'open minded on nuclear'". 12 May. http://news.bbc.co.uk/1/hi/sci/tech/4764933.stm

BBC News (2007). "Nuclear review 'was misleading'". 15 February. http://news.bbc.co.uk/1/hi/uk_politics/6364281.stm

BERR (2007a). *Nuclear Power Generation Cost Benefit Analysis.* http://webarchive.nationalarchives.gov.uk/20100512172052/http://www.berr.gov.uk/files/file39525.pdf

BERR (2007b). *The Future of Nuclear Power: An overview of the consultation process.* January 2008. http://webarchive.nationalarchives.gov.uk/2010 0512172052/http://www.berr.gov.uk/files/file43546.pdf

BERR (2008). *Meeting the Energy Challenge: A white paper on nuclear power.* Cm 7296. January. https://www.gov.uk/government/uploads/system/uploads/attachment_data/file/228944/7296.pdf

Blair, T. (2005). "Speech to Labour party conference". http://www.british

politicalspeech.org/speech-archive.htm?speech=182

CBI (2007). *Climate Change: Everyone's business.* November. http://www.cbi. org.uk/media/1058204/climatereport2007full.pdf

CoRWM (2006). *Managing our Radioactive Waste Safely – Recommendations to government.* CoRWM Doc 700 31 July. http://webarchive. nationalarchives.gov.uk/20100430154107/http://www.corwm.org.uk/ Pages/Current%20Publications/700%20-%20CoRWM%20July%20 2006%20Recommendations%20to%20Government.pdf

DEFRA (2007). *Draft Climate Change Bill.* Cm7040. March. https:// www.gov.uk/government/uploads/system/uploads/attachment_data/ file/229024/7040.pdf

DTI (2006a). *Our Energy Challenge: Securing clean, affordable energy for the long term. Consultation document.* January. http://fire.pppl.gov/ uk_energy_review_2006.pdf

DTI (2006b). *The Energy Challenge: Energy review report.* Cm6887. July. https://www.gov.uk/government/uploads/system/uploads/attachment_ data/file/272376/6887.pdf

DTI (2007a). *Meeting the Energy Challenge: A white paper on energy.* Cm7124. 23 May. https://www.gov.uk/government/uploads/system/uploads/attach ment_data/file/243268/7124.pdf

DTI (2007b). *The Future of Nuclear Power: The role of nuclear power in a low-carbon UK economy.* 23 May. http://130.88.20.21/uknuclear/pdfs/The_ Future_of_Nuclear_Power_Consultation%20Document_May_2007.pdf

EEF (2005). *Sustainable Energy – A long-term strategy for the UK.*

G8 (2005). *Gleneagles Communique on Climate Change, Energy and Sustainable Development and Plan of Action.* http://www.g8.utoronto.ca/ summit/2005gleneagles

Greenpeace (2006). http://www.greenpeace.org.uk/blog/nuclear/greenpeace-launches-legal-challenge-against-the-government

Guardian (2005). "Chief scientist backs nuclear power revival". 21 October. http://www.theguardian.com/science/2005/oct/21/energy.greenpolitics

Guardian (2006). "MPs warn Blair against hasty decision on energy strategy". 10 July. http://www.theguardian.com/environment/2006/jul/10/energy. greenpolitics

House of Commons (2006). *New Nuclear? Examining the issues.* Trade and Industry Select Committee Fourth Report of Session 2005-06. HC1122. July. http://www.publications.parliament.uk/pa/cm200506/cmselect/ cmtrdind/1122/1122.pdf

Institute of Physics (2005). *Press Release: How to plug the energy gap.* 9 November. http://www.eurekalert.org/pub_releases/2005-11/iop-htp 110905.php

IoD (2007). *Press Release: IoD calls for economy-wide application of carbon pricing as best way of tackling climate change.* 19 November. http://

www.iod.com/influencing/press-office/press-releases/iod-calls-for-economywide-application-of-carbon-pricing-as-best-way-of-tackling-climate-change

Nucleonics Week (2006). 6 April.

Platts commodity news (2007). 27 November

Royal Academy of Engineering (2006). *Energy Seminars Report.* http://www.raeng.org.uk/publications/reports/energy-seminars-reports

Stern, N., (2007). *The Economics of Climate Change: The Stern review.* Cambridge: Cambridge University Press.

Telegraph (2005). "Nuclear protester disrupts Blair speech". 30 November. http://www.telegraph.co.uk/news/uknews/1504367/Nuclear-protester-disrupts-Blair-speech.html

Tromans, S. (2010). *Nuclear Law: The law applying to nuclear installations and radioactive substances in its historic context.* London: Bloomsbury.

6 / The sale of British Energy (2006-08)

British Energy returns to the stock market

On 17 January 2005 the shares of the formerly near-bankrupt British Energy (BE) started trading again on the London Stock Exchange, just over two years after the company had been rescued by the government. There was much that was new about the company, including a fresh management team led by an admired Chief Executive, Bill Coley, who had many years' experience in the USA. The company's balance sheet had been cleaned up, with most of its debt replaced by equity, meaning that former creditors now had shares in the company, which they could sell if they wished. The company's formerly inflexible contracts with British Nuclear Fuels Limited (BNFL) for fuel storage and reprocessing, which had played an important part in the company's financial collapse, had been swapped for payments that varied with the financial health of the company, providing a natural cushion against any future fall in power prices.

BE's assets – the old AGRs and the single modern PWR – were, however, the same. Despite new investment and a relentless focus on performance, the older stations were still unreliable and their future was uncertain. The government, through the Nuclear Liabilities Fund (NLF), was a 64% shareholder in the company. The NLF didn't try to influence the company's policy but the ultimate shareholder, the government, had an unavoidable interest in its future, not just financially but strategically.

The year 2006 saw these twin forces pulling in different directions. In April BE announced that it had contracted for two-thirds of the company's output at a selling price more than a fifth above the previous year, which lifted the shares of the company. A rising power price directly raised the value of the company, since its costs were mainly fixed. But in September, news that new cracks had been discovered at two reactors caused a 20% fall in the shares. No matter how high the power price rose the company had to produce the power to benefit from it. And the AGRs could still not be relied upon.

Nonetheless, against the backdrop of a much improved outlook for new nuclear power, the company's value rose. It held a trump card in any future nuclear renaissance – its sites. Anybody wanting to build new nuclear stations in the UK faced the problem of where to put them. Technical and cost arguments and local acceptability made existing nuclear sites the obvious choice and BE owned most of them. The Nuclear Decommissioning Authority (NDA) owned the rest – the Magnox sites.

The older nuclear stations were, if anything, a deterrent to anybody thinking of buying BE. The relatively modern PWR at Sizewell was reliable and used industry-standard technology whereas the AGRs were British and continually troublesome. However, they sat next to land that offered good prospects for building new, state-of-the-art nuclear stations. How those sites could be used was a growing matter of interest to both the government and the BE board.

In May 2007 the government sold 28% of BE to outside investors for £2.34bn, leaving it with 36% and significant influence on the company's future: it was impossible to imagine the company doing a deal against government wishes. But the policy was to leave the board free to run the company while considering how BE could contribute to the government's goal of new nuclear build.

The duty of the BE board was to seek the best value for shareholders. It became increasingly clear to the Finance Director, Stephen Billingham, that this probably meant selling the company. Billingham was typical of finance directors in having an accountancy background but rare in having a PhD in accounting and finance. He had a rich and varied experience in business and had joined BE in 2004, ahead of its new stock market flotation.

Companies often fight to keep their independence, but sometimes that independence is in the best interests only of their senior management. The shareholders may be better off if the company is sold to a bidder for a higher price than the shares are worth in the stock market. A takeover often means senior managers and board members lose their jobs. It is to the credit of the BE board that they pursued their duty to shareholders. (Billingham did lose his job after BE was sold in 2009.)

The argument that BE was worth more to a potential bidder went as follows. Power prices had risen from their lows of the early 2000s, when companies and bankers were all running away from investing in UK electricity generation. There was now a growing need for new capacity because coal stations were closing, and the older nuclear stations were also gradually reaching the ends of their lives. With brisk economic growth, demand was likely to outstrip supply, which implied good investment returns.

This investment value potential was unlikely to be captured directly by BE, though. It lacked financial credibility, as evidenced by its poor credit rating, so it was unlikely that anybody would want to lend to the company except at high interest rates. BE's cash flow was growing rapidly in 2007 on the back of rising power prices, but the company's reputation remained tainted.

Then there was the question of new nuclear potential. When the company got into financial trouble at the end of 2002, the value of new nuclear was entirely notional. Nobody thought that there was money to be made in building new nuclear stations in the UK, and the government appeared unsupportive, if not downright hostile. By 2007, however, government policy had shifted and ministers were now wondering how they could help get the extra investment needed to build new capacity. The government would still not commit taxpayers' funds, but all other policy support was potentially available. With rising fossil fuel prices, supportive policy and optimism about the cost of new nuclear (at least compared with renewables), companies might be willing to buy BE, and pay a premium for its nuclear sites.

Selling BE required at least one bidder interested in new nuclear and with adequate funds. In 2007 there were several potential candidates. The obvious one was the French nuclear giant EDF. But there was also the French company Gaz de France (GDF), formerly just a gas company but now in the process of merging with diversified French energy and utility company, Suez. GDF Suez had a stake in the Belgian nuclear industry and the new group was looking to expand further. The two big German utilities RWE and E.ON both had substantial nuclear operations and large balance sheets. In Spain, Endesa and Iberdrola were both experienced nuclear operators looking for new opportunities. And there was Centrica in the UK, known to the general public as British Gas; it had a large retail business in electricity and gas for which new nuclear investments might be a good fit.

New nuclear investment was appealing to BE staff too. Without new stations, the company was set for managed decline, gradually shrinking as the AGRs closed, leaving only Sizewell B operating well into the middle of the century. It would be hard to retain talented and ambitious people and motivate them to get the best out of the older reactors if they saw no career potential. So BE had a strong interest in trying to get the company involved with new nuclear.

In principle BE could have partnered with another company, sharing its sites in a joint venture with one or more other organizations that would provide cash. Joint ventures can work when there is a roughly equal contribution from the partners and they want to avoid full commitment. But they become unstable if the partners' interests diverge over time; and they also raise management problems, since there has to be agreement between the partners on all key decisions. Joint ventures tend either to evolve into a single merged entity or to fall apart.

Given the unequal financial relationship between BE and any potential partners, it was unlikely that a joint venture could be anything other than a temporary arrangement. In February 2008 BE did invite expressions of interest from potential investors, following the new nuclear white paper that formally gave the government go-ahead for new nuclear. But this was to some extent window dressing. All arguments pointed towards BE finding a company that would want to buy it.

The dance: finding partners

For a company to put itself up for sale is a delicate business. A public announcement can destabilize the workforce, who worry about redundancies and changes to their working conditions. And in any sales situation one doesn't want to appear too keen to sell, and must never seem desperate. The ideal situation for maximizing the value of an asset is to indicate that it could be for sale at the right price and to get the highest number of interested bidders.

BE started the process in the conventional way by asking its investment bank advisers, Rothschild, to sound out potential buyers. Investment banks are the traditional match-makers in corporate deals. They respond to client requests and also actively suggest business deals that they believe will make money. A well-informed banker with active connections across an industry will quickly be able to draw up a list of potentially interested parties, ranked by strategic fit and ability to pay.

BE's market value in January 2007 was about £8bn. A rule of thumb for shareholders selling a company in a takeover situation is to seek a premium of 25-30% to the current share price; this premium reflects the greater value to the new owner of the company being sold (cost savings, new investment potential and so forth). These benefits don't always come to fruition, but BE's existing shareholders would probably not have been willing to sell their shares without a premium of that scale. So BE's board was likely to be looking for around £10-11bn. That was a reasonably sized deal in the European utilities sector. The largest European utilities could plausibly fund an acquisition of that scale, though some of them might have to combine to make it work. In early 2007 French nuclear giant EDF had a market value of about £75bn, and the German utilities E.ON and RWE were worth about £52bn and £37bn, respectively. The largest UK contender was Centrica, worth about £18bn, probably too small to make a bid on its own but enough to make it a potential part of a consortium. So there were plausible bidders for BE.

But would they want BE? In the early 2000s very few utilities, even those with existing nuclear power stations, saw much of a future in new nuclear. However, the combination of climate change policies (particularly the European carbon trading scheme introduced in 2005), rising fossil fuel prices and growing anxieties about the security of Russian gas supplies made nuclear look appealing. And the UK was one of very few European countries with a pro-nuclear government plus a clear need for new power investment. So some of these European utilities might well be interested in BE.

There were also large utilities in the USA, but the outlook for nuclear was deteriorating there because falling natural gas prices made new nuclear look uncompetitive. Nuclear was a large and mostly successful part of the US power market but the major utilities were not looking for overseas ventures, having been disappointed by previous foreign investments, including in the UK. So the field was limited to Europe.

Rothschild discreetly sounded out the market, mindful of the board's desire not to cause any publicity that might interfere with the May 2007 white paper on energy and the accompanying new nuclear consultation paper. The white paper said the government had reached the "preliminary view that it is in the public interest to give the private sector the option of investing in new nuclear power stations". After Greenpeace's successful legal challenge to the 2006 decision to open the nuclear option, this time everything needed to go smoothly.

Things started out well. In late 2007 BE held talks with most of the large European utilities: Centrica, EDF, GDF Suez, E.ON, RWE (the last in partnership with Swedish electricity company Vattenfall) and Iberdrola. The only two missing were the other major Spanish utility, Endesa, and the Italian power company Enel (which was in advanced takeover talks with Endesa).

It costs relatively little to express interest, and in a major transaction like the sale of BE, many companies would want to find out as much as possible, keeping their options open without committing themselves. Nevertheless, this was a promising start in terms of creating competition that would maximize the selling price.

The government, which was the largest single shareholder, had a wider set of goals. The government-owned shares were managed by the Shareholder Executive, an agency set up in 2003 as part of the Department of Business, Enterprise and Regulatory Reform (BERR) (later replaced by the Department of Business, Innovation and Skills (BIS)) to manage the government's various shareholdings on a commercial and professional basis. It was staffed with former City people who brought commercial and financial knowledge to help the government get a better deal when negotiating with private sector partners or bidders. The Shareholder Executive already had the job of monitoring BE's performance and was the main adviser on the actual sale.

Price was *not* the main consideration. The government's primary objectives (later spelled out in a National Audit Office (NAO) report on the eventual sale) were:

- ensuring that nuclear operators are able to build and operate new nuclear power stations from the earliest possible date, to the widest possible extent, with no public subsidy, and with all unnecessary obstacles removed;
- and maintaining the viability; and continued safe operation of the existing British Energy fleet.

The secondary objectives were to:

- minimise the Government's exposure to risk of being unduly dependent on a single company for nuclear going ahead in a timely fashion; and
- maximise the value of the Government's interest in British Energy and Nuclear Decommissioning Authority sites.

(NAO (2010) para 1.7)

Maximizing current and future nuclear output was what mattered most to the government, and ideally with more than one company to provide some competition. Value for taxpayers came second.

The reference to Nuclear Decommissioning Authority (NDA) sites was to the ownership of the old Magnox stations. Most of these had closed and the rest were scheduled to close by 2014, but the sites were potentially valuable for new build. The NDA, which took over most of the operations of the former BNFL, was owned by the government. Its main job was and remains cleaning up sites contaminated by radiation and other hazardous nuclear materials at the Magnoxes and AGRs, plus managing the sprawling mess of reprocessing and other facilities at Sellafield.

The government set up a Nuclear Sites Steering Group, whose members were drawn from the Shareholder Executive, BERR and the Treasury. The steering group was to oversee the Shareholder Executive team of former mergers and acquisitions (M&A) bankers who represented the government in the BE sale process.

The formal process of selling BE started in January 2008. The timing looked auspicious. The value of BE was underscored by the publication at the same time of the new white paper on nuclear power, which gave the formal go-ahead to new nuclear power (Chapter 5).

The new Committee on Climate Change (CCC), chaired by Lord Adair Turner, also started in January. The second stage of the European Emissions Trading System (ETS) began, with Norway, Iceland and Liechtenstein joining.

One by one, however, the bidders for BE began to drop out. E.ON had decided to buy a large chunk of Spanish electricity assets as part of the takeover of Endesa by the Italian utility Enel, so it now lacked the financial resources to buy BE. Iberdrola and GDF Suez both decided the deal wasn't right for them. In February Centrica made an indicative offer but wanted to pay with shares rather than cash. This didn't meet the goals of the private shareholders or the government, so Centrica was dropped from the process.

By March there were only two possible bidders left, EDF and RWE. Both were given access to BE's "data room", meaning they could examine BE's books in much more detail than before, in exchange for signing confidentiality agreements, should they later leave the process. On 17 March BE told the stock market that it was in talks that might lead to a bid. RWE had put in an indicative offer of around 700p.

Founded in 1898, RWE was a diversified energy company based

in the industrial city of Essen in the Ruhr area of North Rhine-Westphalia. Its roots were in its name, which stood for Rheinisch-Westfälisches Elektrizitätswerk. With around 20 million customers it was a big company, with access to bank financing large enough to fund the bid, but it was a very large deal even so. RWE was keen to bring in a partner to share the risks and the funding.

Centrica, having decided it wasn't financially strong enough to mount a bid on its own, was flirting with both RWE and EDF as possible partners, hoping that bringing a British element to an otherwise continental European field might help sway the British government. It was also a big buyer of electricity, which had some value in the deal. But Centrica failed to come to any definite arrangement with either company.

In late March the British press reported that Vattenfall of Sweden was teaming up with RWE to make a joint bid. Vattenfall – "waterfall" in Swedish, reflecting its origins as a hydroelectric generator – was fully owned by the Swedish government. The company, under its CEO, Lars Josephsson, had made a name as a leader in the fight against climate change, even though some of its foreign investments included brown coal power stations, among the most polluting. With seven nuclear power stations in Sweden, expanding into UK nuclear was consistent with Josephsson's corporate strategy.

The Swedish government had a long-standing tradition of allowing state-owned companies to make their own decisions. But buying into a major British nuclear power company was something out of the usual, even though Sweden had nuclear power and Vattenfall had many other foreign investments in Germany, Poland, Denmark and Finland. It rapidly became clear that the Swedish government, a coalition with divided views on nuclear power, was not happy about Vattenfall's plans. On 1 May 2008 Josephsson had to make an embarrassing phone call to his counterpart at RWE, Dr Jürgen Grossmann, to tell him that the deal was off. RWE initially said it would continue on its own, but the deal was too big and risky and RWE then withdrew, leaving only EDF in the bidding.

EDF is not just any company. It is the French equivalent of the CEGB. EDF has managed to remain the dominant generator and supplier of power in France and is still 84% owned by the state. The reason it is the world's largest nuclear generator is that France lacks indigenous fuel resources.

In the 1950s and early 1960s France followed the same reactor technology path as the UK, and for similar reasons (lack of access to enriched uranium and to heavy water). However, unlike the UK, France dropped that technology in the late 1960s (when the fiercely nationalist former President de Gaulle could no longer object). Instead it bought the emerging global standard PWR from the US company, Westinghouse but paid to license the technology for its own use and development. After the Arab oil embargo in 1974 showed the danger of being dependent on other countries for imported energy, France (for which oil represented two-thirds of its energy) over the following eight years ordered forty reactors; by 1985 two-thirds of French electricity was nuclear. This was the largest expansion of nuclear power in history, an effort that is only now being surpassed, by China.

EDF had the nuclear expertise and the backing of a state. The French government only reluctantly floated the company on the stock market in November 2005, selling a small stake to the private sector. EDF is now something of a hybrid of state and private company. It has shareholders, it follows normal corporate governance and sets its own strategy. Yet nobody should be in doubt that it is a creature of the state; after all, French law requires the state to maintain at least a 70% stake in EDF.

For a few weeks it looked as if even EDF might also withdraw. The purchase by Italian electricity company Enel of the Spanish utility Endesa had encouraged speculation about further acquisitions. EDF was reported to be in talks with the Spanish construction company ACS about a joint bid for the other major electricity company, Iberdrola, and possibly the third, Union Fenosa. ACS already had a 45% stake in Fenosa and 12% in Iberdrola, so having a Spanish shareholder might soften the politics of Spain losing the bulk of its power sector to foreign owners.

Iberdrola made it clear it would fight "to the death" for its independence, a novel tactic in takeovers but one with local political support. One group in EDF's management team saw a major investment in Spain as the logical strategy for EDF, given the two countries' interconnected power systems. But an opposing group argued, eventually decisively, that EDF's core interest was nuclear and that a bid for BE would maximize the potential for the company's long-term expansion. Given that EDF, large as it was, couldn't finance acquisitions in both Spain *and* the UK, it was lucky for BE and the British government that the pro-UK faction eventually won, leaving EDF still in the running for BE. Fenosa eventually did a deal with Spanish company Gas Natural instead.

The game of poker

From the seller's point of view, having a single buyer is, to say the least, not ideal. However, BE had put itself up for sale and it would be embarrassing if it failed to find a buyer. EDF, knowing that all other possible buyers had dropped out, would now be able to pay less.

Valuing any company is a mixture of science and judgement. Since the future is unknowable and most of the value of a company lies in the future, there is room for reasonable disagreement about what a company is worth. That is especially true when a major reason for buying the company is as a platform for building new power stations that might in fact never be constructed.

BE didn't have to sell itself at any price. It could return to being an independent, standalone company, operating its ageing power stations as best it could and maximizing the cash flow it could then pay to shareholders. Not an exciting future but perfectly feasible. But that wouldn't give the government what it wanted – a credible new nuclear investor. So the government and the private shareholder interests were not entirely aligned at this point.

BE's board had set a deadline for formal bids. On the day of the deadline expiring, 1 May 2008, EDF put in an offer of 705p/share. That compared with BE's share price of 721p, which reflected market optimism about how high a bid might go. How to judge whether that was a good price?

The Nuclear Sites Steering Group, which coordinated the various government entities interested in the outcome of the BE deal, had in March 2008 appointed the Swiss bank UBS as financial adviser, selecting it from three banks that tendered for the business. This was quite separate from BE's own use of Rothschild as an investment bank adviser. UBS produced a set of valuations that were later made public in the NAO report.

A company valuation depends on having a sensible method or process and a set of assumptions. The method is relatively straightforward and widely followed: you forecast the company's cash flows into the future, perhaps 20 or 30 years ahead, and discount those cash flows at an interest rate that reduces the value of future cash flows relative to current ones. The choice of discount rate can be somewhat controversial, but what really drives the valuation is the choice of future cash flows.

UBS made assumptions about two sets of things. The first were loosely under the company's control, including output and costs. But it then had to forecast things that not only were outside the company's control but were inherently unpredictable. Chief of these was the future price of electricity, which in turn depended heavily on fossil fuel prices, led by oil. Oil prices are notoriously unpredictable in the short term, let alone more than a decade in the future. UBS used a range of $102 to $109 a barrel of oil as the key assumption.

Another variable was the price of carbon in the European ETS. The higher this price the more valuable zero-carbon electricity would be, since it was exempt from paying the carbon price. UBS used a figure of €30/t.

These assumptions, together with the company's balance sheet, led UBS to value the company's existing nuclear business at 601p, plus 33p for the Eggborough coal power station.

The other factor was the value of the new nuclear business potential. That was also a matter of judgement, since it involved a set of conjectures about how many stations might be built, and when, and what the value of BE's sites would be compared with other possibilities, such as those owned by the NDA. Somehow UBS settled on a figure of 100p for new nuclear, for a total valuation of 734p/share.

If this was approximately correct, then EDF's bid of 705p was too low and the government was disinclined to accept the offer. For its part the BE board had already formed a strong view that the offer was too low and therefore rejected it.

This stage in a takeover, particularly with a single bidder, starts to resemble a game of poker. BE's board faced the risk that EDF, having been rebuffed, would walk away. BE may have believed, and certainly needed to seem to believe, that EDF was just trying it on with the first bid and was sufficiently interested to put it a better bid, having spent time, effort and public reputation in getting this far. But there was a risk: if EDF really did walk away then the shares could drop back to their value without any new nuclear, nearly 100p lower.

From EDF's point of view, the company knew it was the only bidder so it faced no pressure to outbid an overenthusiastic rival. It wanted BE; its plans for new nuclear investment in the UK would be very difficult to carry out without it. It might still have gambled that the government's need for nuclear investment would somehow lead to a deal in future, but it knew that the British government had a record of not overriding shareholder interests just to achieve its policy goals.

Takeover bids are rather like a man asking a woman to dance in a Jane Austen novel. The bidder's initial offer is rejected but not too rudely, so as to leave the door open to a second, improved offer. That offer is then accepted, or perhaps forms the basis for an even better final offer, which arises from close negotiation. The deal is done and everyone is happy.

The government expected something like this to happen and at first it seemed to be right. In July EDF came back with an offer of 765p a share, about 9% above its May bid. The government, advised by UBS that the value of BE was probably around 734p, was happy to accept this. The BE board's first inclination was to accept it too. A positive and encouraging message went back to EDF that the deal was probably going to happen. The British ambassador in Paris informally sounded out EDF's main shareholder – the French government – to check. The French equivalent of the Shareholder Executive indicated that it would of course block EDF from any overbidding but it seemed relaxed about the deal on the table. There was a wider sense of optimism that the French and British could collaborate on new nuclear power in ways that would benefit both countries. Indeed, some City analysts believed that the French government, actively promoting the interests of the French nuclear industry, was the main force behind the bid in the first place. EDF, confident that the deal was going ahead, booked a set of rooms on the Champs-Élysées and sent out invitations to journalists for a major press conference.

But BE's board had to check with its main private shareholders. The board could only recommend the offer; the decision rested with those who owned the shares. The British government had indicated it was happy to accept the offer on the table. Two major private shareholders made it clear that they were not.

M&G and Invesco Perpetual were two large retail fund management companies that held between them 22% of BE's shares, enough to damage and probably block the deal. No board would want to push a deal that more than a fifth of shareholders opposed. Both shareholders were concerned that BE was still selling itself too cheaply, partly because they were optimistic about future power prices and about BE's future output. These were both unknown factors but they had a big impact on the value of the company, as UBS's valuation work for the government had confirmed. M&G and Invesco Perpetual were both doing what they saw as their job, getting the best value for their clients, whose money they had invested in the shares of BE. But they had no particular interest in the strategic value of the deal for British nuclear policy, unless that could be crystallized as cash for their clients.

BE's board reluctantly told the board of EDF that it could not recommend the bid and the press conference was cancelled. BE's shareholders were now playing quite a risky game. There was less reason to think that EDF would come back with an even better offer and it might well have concluded that the shareholders were unreasonable and impossible to satisfy without paying too much.

The British government, having accepted the deal, was dismayed. Just when it seemed to have secured both a good price for the taxpayer and, more importantly, the entry of EDF into the British nuclear industry, the deal had been wrecked by two City investors.

The various advisers to BE and the government were all highly motivated to find a way to rescue the deal, since the convention in M&A advice is that payment depends on success. No matter how many hours of analysis the bankers at Rothschild and UBS put in, they risked getting only a token fee if no deal took place. They duly came up with an innovative solution. Normally a takeover bid is made either in cash or in the shares of the bidding company. Cash is always better since it is the more flexible for the seller, who can always buy shares in the new combined company if desired. EDF was financially strong enough to offer cash but the private shareholders thought it too little.

So a new form of payment was created – the "nuclear note". This was a derivative contract, a financial instrument whose value depended on the future combination of the wholesale price of electricity and the output achieved by BE.[1] The value of a share in BE was highly dependent on both of these factors, as well as on a range of other matters, including the general level of share prices and economic confidence. The "nuclear note" was more narrowly designed, to track prices and output. The deal was changed to allow shareholders to sell some of their shares for the nuclear note instead of cash. That way, if power prices and output did indeed do better than expected, those shareholders would benefit, and would not feel they had sold their shares too cheaply. But it was not a guaranteed bet. The value of the nuclear notes could go down, and under some power price and output scenarios would be worthless: For example, if output fell below 50 TWh and the wholesale power price fell to £45/MWh or less then the nuclear notes would pay nothing (NAO, 2010).

---

[1]   The instrument was technically a form of contingent value right (CVR), a device used to bridge major differences between a buyer and seller in a deal, usually because of a specific event. These devices are fairly rare in UK deals.

By September a new deal seemed possible. The nuclear note had won the private shareholders round. EDF added an extra 9p a share to make a final offer of 774p/share. Or shareholders could take 700p in cash plus the nuclear note. The government wouldn't take this option, since it wanted the certainty of cash, which would be put into the Nuclear Liabilities Fund (NLF).

This deal, the shareholders told the board, they could accept. The BE board duly recommended the offer and EDF was able to announce, finally, on 24 September that it was buying BE, subject to approval from the European Commission (EC). Adrian Montague, BE's Chairman, told the press the deal was "good for shareholders, good for staff, good for the nuclear industry and good for the country". This is the sort of statement often made at the time of large corporate deals. For once it was probably true.

Then Centrica announced that it was negotiating to buy 25% of BE from EDF plus an option on investing in future new nuclear stations. Having nearly bid, nearly partnered with a number of companies and irritated quite a few of them, Centrica was back in the game.

As a condition of buying BE, EDF agreed to sell at least two of its nuclear sites to third parties, keeping the option open for other investors in new nuclear power. So the government achieved its goals of pushing new nuclear without being entirely dependent on a single company.

The private shareholders who held out for a higher value got their wish but those who opted for the nuclear note instead were out of pocket in January 2010, when the nuclear notes were trading at 35p. So they had received 700p cash plus an instrument worth 35p for a total of 735p; the government, having opted for the full cash offer, had banked 774p. As a rough comparison with what might have happened to BE shares if EDF had walked away, the share price of the coal power generator Drax Group had nearly halved since EDF bought BE. Of course, EDF may have placed a much higher value on the new nuclear potential than the 100p/share that UBS assumed. EDF seemed happy with the deal and had much bigger things in its sights than the short-term price of electricity.

The deal was still not completely out of the woods, though. Given its scale and potential impact on the UK power market, it was referred to the EC for approval. The commission was not concerned with the matter of new nuclear stations, which it would have a chance to consider many years later. But it did worry that the combination of

EDF's existing power generation and retail business with BE would lead to excessive concentration in the market, damaging competition. It also noted that the company would own three grid supply points connected to National Grid's transmission network at Hinkley Point, which would stop anybody else from building a power station at that site, whether nuclear or not.

So the commission approved the deal in December 2008, but with conditions. EDF had already promised to sell a site adjacent to power stations at either Dungeness in Kent or Heysham in Lancashire, if it got permission to build a new nuclear station anywhere in the UK. The commission made this sale unconditional, so that even if EDF didn't go ahead with new nuclear investments, another company would have a chance. The commission also required EDF to sell its Sutton Bridge gas power station, its Eggborough coal power station and one of the three Hinkley Point grid connections. Even after the disposals EDF would end up with about 17% of the total power generation market (the next largest was RWE's npower subsidiary with 12%).

On 13 January 2009 EDF completed the purchase of BE. The proceeds of the government's stake, £4.4bn, were paid into the Nuclear Liabilities Fund, which at £8.3bn was for the first time comfortably more than the expected cost of decommissioning BE's nuclear power stations, a cost the company had passed to the government as part of its financial restructuring in 2003.

In May 2009 Centrica confirmed that it was buying part of BE from EDF; however, it was acquiring only 20% and for less cash than originally discussed, reflecting the fall in UK power prices since the previous autumn. It would also have the option to take part in new nuclear investment, if and when that actually happened.

Conclusion

Having sold BE to the best possible buyer, the government's policy was making progress. The sale was completed shortly after Gazprom announced on 1 January 2009 that it was again cutting gas deliveries to Ukraine. What clearer warning could there be of Europe's dependency on foreign energy? Nuclear offered low-carbon power, security of supply and economical electricity. Unfortunately, the global financial crisis was about to undermine the economic case for nuclear.

REFERENCE

NAO (2010). *The Sale of the Government's interest in British Energy.* HC 215 Session 2009-10. 22 January. http://www.nao.org.uk/wp-content/uploads/2010/01/0910215.pdf

7
The Climate Change Act strengthens the case for nuclear (2008)

The Committee on Climate Change spells things out

The scale of the challenge to meet the Climate Change Act goals was made clear by the Committee on Climate Change (CCC) in its first report, published in December 2008. It showed the difficulty of cutting emissions by 77% by 2050 compared to 2006 levels (equivalent to a cut of 80% from 1990 levels). As Figure 7.1 shows, CO_2 emissions mainly come from areas other than electricity generation, so there was a need for a radical change in energy use across the economy.

Figure 7.1: The 2050 carbon emission goal versus 2006 actual emissions (millions of tonnes of CO_2 equivalent)

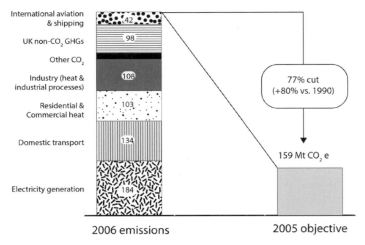

Source: CCC (2008) Fig. 2.1

102

The best, probably only, way to achieve such an ambitious goal was to decarbonize transport by converting it to electricity. The same might be true to some extent for heating, another major source of carbon emissions. The electricity could then be sourced in a low-carbon way. But that meant a rise in total electricity generation. Finding very large quantities of low-carbon electricity became correspondingly important.

The CCC then offered a possible "pathway" to the 2050 emission target, taking into account the fact that international aviation and shipping emissions were not covered by the government's policies and that some non-CO_2 emissions would be very hard to cut (the ultimate goal is to cut greenhouse gas (GHG) emissions, of which CO_2 are the largest part). The upshot was that areas where policy *could* be applied were under even more pressure (see Figure 7.2).

Figure 7.2: Committee on Climate Change projected UK carbon emissions pathway (millions of tonnes of CO_2 equivalent)

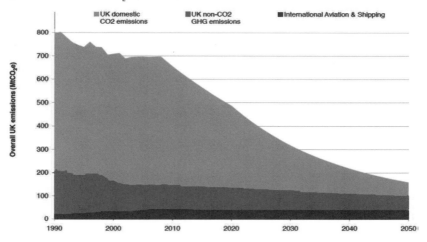

Source: CCC (2008) Fig. 2.28

To keep to the Figure 7.2 pathway, electricity would need to be almost completely decarbonized by 2030, only 21 years from the date of the report's publication at the end of 2008. Specifically the carbon content would need to be cut from a forecast 500 g of CO_2 equivalent per kWh of energy in 2010 to less than 70 g in 2030.

That in turn could be achieved, the CCC argued, by an expansion of nuclear, renewables (mainly wind) and by using carbon capture and storage (CCS) to reduce the emissions from the remaining fossil fuel power stations that would be needed to manage the balance between demand and variable supply (at night wind power supply is lower, while solar supply is zero).

Using low-carbon electricity as a way to cut wider economy carbon emissions would entail a rise in annual electricity generation to around 570 TWh in 2050 compared with about 440 TWh with no emissions targets.

How would this power be generated without raising carbon emissions? Much depended on CCS, which the CCC admitted was "currently not a proven technology at full commercial scale". CCS – extracting most of the carbon dioxide created by burning fossil fuel and safely storing it – could allow some coal and gas stations to run while keeping emissions very low. But if CCS turned out not to be available, then a "huge" expansion of nuclear would be needed. If nuclear were also ruled out, then renewables' share of total electricity generated would have to rise to 60%; this would be "at substantial additional cost" and "with greater energy demand reduction (either price-induced or via life-style change)."

A more detailed projection of the central case scenario's electricity source split was published in a report written for the CCC by consultants MARKAL (MARKAL-MED (2008) Fig. 2.14). MARKAL's scenario showed nuclear electricity in 2050 representing nearly double its 2000 level and wind expanding to a quarter of total UK electricity. However, Markal's numbers only met the 2050 target because they assumed a large use of CCS in expanded coal-powered generation plus a smaller amount used in gas-fired generation. This otherwise plausible scenario depended on a technology that was not yet available and might never be so.

A Cambridge engineer makes things clear

In 2009 Professor David MacKay of the University of Cambridge published a book that laid out with unprecedented clarity the logic of British energy options. *Sustainable Energy – Without the hot air* used a target for UK energy consumption of 125 kWh per head per day as its central reference point. This was the average consumption in 2008, but excluded energy embedded in goods imported from abroad (MacKay, 2009). MacKay's book showed how decarbonization could be reconciled with British energy needs. Most importantly, using highly effective graphs and diagrams, he contrasted what was physically feasible with what was practical and realistic.

For example, a plausible (but extreme) maximum for deploying onshore wind energy might be to cover 10% of the UK with turbines (mostly in the north and west where wind speeds are much higher). That would generate the equivalent of 20 kWh/day per person, about half the amount consumed by an average fossil-fuelled car driven for 50 km. This would be an enormous expansion of wind power, representing around seven times the existing wind capacity already installed in Germany, which was far ahead of the UK in wind at that point. (Outside the scope of MacKay's book was the consideration that it would also mean covering many areas of outstanding natural beauty with intrusive wind farms, connected with equally intrusive transmission wires. And much of the capacity would be Scotland, which could not be depended on to remain part of the UK.)

MacKay then considered offshore wind in shallow waters, already under development and less controversial than onshore wind. He estimated that the total realistic supply would be 16 kWh/day/person, and that would involve covering a sea area two-thirds the size of Wales, the equivalent of constructing a strip of wind turbines 4 km deep along the entire UK coastline. Conceivably a greater area could be covered but there would be conflicts with fishing, shipping and other environmental priorities.

Deep offshore wind, still at the experimental stage, could potentially add a further 32 kWh/day/person, if it was possible (and economic) to put a strip of turbines 9 km deep around the entire UK coastline, or the equivalent area.

On solar, MacKay estimated a large-scale application of rooftop solar PV (photovoltaics) (as opposed to the cheaper solar heating of water) could bring 5 kWh/day; "large-scale" meant covering every south-facing roof in the country. Solar parks that take up other land (which of course means displacing some other use) might generate 50 kWh/day/person, but that would take up 5% of the UK's surface area (and recall that we have already earmarked 10% for wind). The 50 kWh/day/person figure assumed improved technology that would allow the UK to get 10 W of power per square metre of land covered, which was double the rate achieved by an actual solar park in Bavaria, Germany.

One of the most useful things about MacKay's book is that it emphasizes physical limits, even before we consider the economic limits, which would make many of the physical possibilities impractical on cost grounds. In particular, a small, crowded island like the UK has to consider energy intensity per square metre. This is the dimension in which nuclear scores highly. Wind generates about 2 W of power per m^2, solar PV about 22 W, and nuclear 1,000 W.

The same space problem arises with biomass, which theoretically could be used to generate 24 kWh/day/person if we give up another chunk of the UK's land area. MacKay estimated that tidal power, still in development, might yield 11 kWh/day/person, and wave energy, even less proven, some 4 kWh/day/person. Together with solar heating, a large proportionate expansion of the currently very small amount of existing hydro (which is very hard to expand, both practically and politically) and a very small amount of geothermal, MacKay suggested the UK might just about stretch to a renewable energy total of 178 kWh/day/person.

This is well above the daily average consumption of 125 kWh/day so at first it seems the UK could rely entirely on renewables. But as a reality check MacKay compared his figures with those from various other reports on renewable energy potential in the UK: the Institute of Electrical Engineers (IEE), the Tyndall Centre, the Interdepartmental Analysts Group (IAG), the Performance and Innovation Unit (PIU) of the UK government and the Centre for Alternative Technology (CAT) (Figure 7.3).

Figure 7.3: Comparison of estimated potential energy from renewable UK sources (kWh/day/person)

Renewable energy source	Study					
	MacKay	**IEE**	**Tyndall**	**IAG**	**PIU**	**CAT**
Geothermal	1.0	10.0				
Tide	11.0	2.4	3.9	0.1	3.9	3.4
Wave	4.0	2.3	2.4	1.5	2.4	11.4
Deep offshore wind	32.0					
Shallow offshore wind	16.0	6.4	4.6	4.6	4.6	21.0
Hydro	1.5		0.1			0.5
Biomass	24.0	4.0	2.0	3.0	31.0	8.0
Solar PV farm	50.0					
Solar PV	5.0		0.3	0.0	12.0	1.4
Solar heating	13.0					1.3
Wind	20.0	2.0	2.6	2.6	2.5	1.0
Total	178	27	16	12	56	48
Average UK consumption	125	125	125	125	125	125

Source: MacKay (2009) Fig. 18.6

Figure 7.3 shows that MacKay's estimates are for the most part much higher than the others, so his estimates of energy potential should be considered highly optimistic. No other source includes much for deep offshore wind and MacKay has far higher estimates for solar and onshore wind. MacKay intended his comparisons to push home the message that while it was *physically* possible (without any economic, political or aesthetic considerations) to achieve full renewable energy independence for the UK, it was not a realistic prospect.

What *is* realistic? MacKay's personal view (at the time of the book) was that, once we rule out the various practical objections to his earlier physical scenarios, we might only get 18 kWh/day/person from renewable sources. Even if we assume more optimistic efficiency of solar PV and rapid progress on tidal, wave and deep offshore wind, this is nowhere near the UK's 125 kWh/day/person average consumption.

MacKay then illustrated some future scenarios in which we cut the average consumption to 68 kWh/day through hugely improved energy efficiency. That included 48 kWh/day/person of electrical energy, which would be used to power transport other than air and to provide a big fraction of heating. But 48 kWh/day/person represents a tripling of average electricity consumption from 2006 levels.

The plans are summarized in Figure 7.4, which illustrates a range of energy strategies, including one without any nuclear and one with a great deal. The plans are D (diverse option), N ("nimby" ("Not In My Back Yard")), L (no nuclear), G (no nuclear or coal) and E (economic). Note that nuclear makes its largest appearance in the economic plan. The nimby plan envisages heavy reliance on solar in deserts (see below) and 72% of UK electricity being imported. There is still 10 kWh/day/person of nuclear in this scenario, requiring 25 GW of nuclear capacity, compared with around 9 GW in 2015.

A crucial component of two of the plans is imported electricity from large-scale solar generation in the deserts of North Africa, something that is currently barely even under consideration and most unlikely to be acceptable to European governments for the foreseeable future. Another important fraction comes from clean coal, meaning CCS fitted to existing or new coal power stations.

If we drop the desert solar component, plans N and L no longer work. If we scale back the amount of CCS that can be fitted (given that it remains commercially unproven), we are left with a hole in all but plans G and E. Plan G depends on a massive expansion of wind and an equivalently large form of storage to balance out the variation

Figure 7.4: MacKay's five energy plans for the UK (kWh/d)

plan D	plan N	plan L	plan G	plan E
Clean coal: 16 kWh/d	Solar in deserts: 20 kWh/d	Solar in deserts: 16 kWh/d	Solar in deserts: 7	Nuclear: 44 kWh/d
Nuclear: 16 kWh/d			Tide: 3.7	
	Clean coal: 16 kWh/d	Clean coal: 16 kWh/d	Wave: 3	
Tide: 3.7			Hydro: 0.2	
Wave: 2	Nuclear: 10 kWh/d	Tide: 3.7	Waste: 1.1	
Hydro: 0.2		Wave: 2		
Waste: 1.1	Tide: 1 kWh/d	Hydro: 0.2	Pumped heat: 12 kWh/d	
	Hydro: 0.2 kWh/d	Waste: 1.1		Tide: 0.7
Pumped heat: 12 kWh/d	Waste: 1.1 kWh/d	Pumped heat: 12 kWh/d	Wood: 5 kWh/d	Hydro: 0.2
	Pumped heat: 12 kWh/d		Solar HW: 1	Waste: 1.1
Wood: 5 kWh/d		Wood: 5 kWh/d	Biofuels: 2	
Solar HW: 1	Wood: 5 kWh/d	Solar HW: 1	PV: 3	Pumped heat: 12 kWh/d
Biofuels: 2	Solar HW: 1 kWh/d	Biofuels: 2		
PV: 3 kWh/d	Biofuels: 2 kWh/d	PV: 3	Wind: 32	Wood: 5 kWh/d
Wind: 8 kWh/d	Wind: 2 kWh/d	Wind: 8		Solar HW: 1
				Biofuels: 2
				Wind: 4

Source: MacKay, 2009: Fig. 27.9. D: diverse option; N: "nimby"; L: no nuclear; G: no nuclear or coal; E: most economic

in wind supply compared with more steady electric demand. MacKay suggested an expansion of pumped storage, an existing technology in which off-peak cheap electricity (probably free if it's from wind) is used to pump water uphill. The water is then released downhill through a turbine when it's needed. The UK has such a facility, a 1.7 GW power station at Dinorwig in Snowdonia, Wales. Dinorwig is not visible because most of it is inside the Elidir Fawr mountain. Originally built to provide back-up in case of a sudden loss of power from a nuclear power station, Dinorwig is used both for balancing out daily demand/supply variation and for short-term emergency extra power. Plan G requires the equivalent of 400 Dinorwigs to be built. That is a lot of mountains that would need to be hollowed out, or as MacKay suggested, 100 of the UK's lakes and lochs would need to be converted to hydroelectric generation.

Conclusion

The CCC reports and David MacKay's book showed it would be very difficult and probably very expensive to achieve the 2050 goals, and even harder if CCS technology wasn't available. An aggressive expansion of wind would not be enough because there would be popular opposition both to the physical scale of the windmills needed and to the expansion of pumped storage to balance the system. It was hard to avoid concluding that the Climate Change Act had committed the UK to a large expansion of nuclear power. That was not what many of its supporters had intended.

REFERENCES

CCC (2008) *Building a low-carbon economy – the UK's contribution to tackling climate change*. December www.theccc.org.uk/publication/building-a-low-carbon-economy-the-uks-contribution-to-tackling-climate-change-2/

MacKay, D. (2009). *Sustainable Energy – Without the hot air*. UIT Cambridge. www.withouthotair.com

MARKAL-MED (2008). *Model Runs Of Long Term Carbon Reduction Targets in the UK Phase 1*. November http://archive.theccc.org.uk/aws3/MARKAL-MED%20model%20runs%20of%20long%20term%20carbon%20reduction%20targets%20in%20the%20UK%20-%20AEA%20-%20Phase%201%20report.pdf

PART III

Turning targets
into action
(2009-13)

8

Making it all happen
(2009)

Progress along the nuclear pathway

The energy white paper of January 2008, which officially reintroduced new nuclear build to UK energy policy, included an indicative pathway by which new nuclear build might start in 2013. For this to happen, the government needed to do a number of things:

- Complete the new generic design approval process for the types of reactor that might be built;
- Reform the national planning system to avoid each new nuclear station having to go through a long, costly and repetitive inquiry;
- Identify suitable sites, with appropriate consultation;
- Put in place a policy to handle and ensure the funding of nuclear waste and the decommissioning of power stations.

Generic design approval

The UK's dismal history of ever-evolving designs had left a legacy of non-standard reactor models and bespoke engineering solutions. The generic design approval (GDA) process was intended to force engineers to produce a finished design that could then be checked and, if approved, could not be changed. While that might lose some modest benefits in incremental design features, that cost would be more than offset by having a fleet of identical reactors. Once the first was approved, subsequent identical reactors would automatically be approved. The process was to be detailed and thorough at the start; swift and straightforward in future.

The ideal of a common design fleet was in conflict with another

113

goal of policy – to create competition between different reactor vendors, to push prices down. Ideally there would be an auction among competing approved designs. But that would be hard to do without slowing down the process. Two reactor types started the GDA process in 2009: the French Areva EPR and the US–Japanese Westinghouse AP1000.

The Health and Safety Executive's (HSE's) first GDA progress report, for January–March 2009, confirmed that it was on track to complete the GDA for the EPR and AP1000 reactors by the target date of June 2011.

Planning and sites

Progress on planning had already started with the November 2007 planning bill. The reason for changing the planning system was given in the 2008 energy white paper. The planning process for nuclear in the past had been "inefficient, costly and lengthy, and in some cases may not have provided sufficient opportunity for consideration of local issues because they spent too much of their time dealing with broader national issues. For example, the nuclear power station Sizewell B took six years to secure planning consent, costing £30m, and only 20 of the 340 inquiry days were devoted to local issues" (BERR (2008) 3.11).

The planning reforms would introduce a new Infrastructure Planning Commission (IPC) to make decisions on strategically important investments. The goal was to have as much pre-agreed decision-making as possible to streamline the process for new nuclear approval and stop the risk of unpredictable appeals that could block new stations late in the process resulting, in very high costs. Instead of each inquiry revisiting the general argument for nuclear power and giving opponents time to restate their hostility, a higher-level argument for nuclear would be embodied in the planning statement, and individual inquiries would focus on local issues.

The policy would be conducted through the publication of a set of National Planning Statements (NPSs), a form of words not heard in the UK since the 1960s. There would be an overarching statement, and individual statements for each key area of infrastructure, including for nuclear energy. Once the NPS had been consulted on and finalized, the planning process would then consider only local issues, avoiding the risk of repetitive inquiries for each new nuclear station.

Central to planning was identifying sites for the new stations. The government published a process for consultation – the Strategic

Siting Assessment (SSA) – in July 2008. This was a detailed, 90-page document, with an extensive literature review. The seriousness of the consultation could hardly be doubted. The purpose of the SSA was to identify sites which "are strategically suitable for deployment of new nuclear power stations by the end of 2025".

The consultation involved 89 written submissions and 3 stakeholder events (in London, Bristol and Manchester), which more than 100 people attended. In its response in January 2009, the government announced that nominations for new sites were invited by 31 March 2009. The SSA would then choose from these sites using strategic criteria. The new IPC need not then consider these issues further, only reviewing specific proposals to build on each site. The process covered England and Wales only.

Eleven sites were nominated and subsequently assessed by the government in line with the various criteria published in the SSA. EDF Energy announced that it intended to build four new EPR nuclear reactors in the UK (with Centrica, which had bought 20% of British Energy (BE) from EDF and an option for a 20% stake in EDF's new build plans for a total of £2.3bn); the first was to be operational by the end of 2017. RWE and E.ON formed a 50/50 joint venture, Horizon Power, in January 2009 with the intention of building 6,000 MW of new capacity at Oldbury and Wylfa. The list of other sites nominated included Bradwell, Braystones, Hartlepool, Heysham, Kirksanton and Sellafield. Of these, only Braystones and Kirksanton were not on existing nuclear sites and both were later dropped.

One other existing nuclear site, Dungeness in Kent, was later dropped because the expected local environmental damage would be too high.

Three of the nominated sites, being attached to old Magnox nuclear stations, belonged to the Nuclear Decommissioning Authority (NDA). These were Oldbury, Wylfa and Bradwell. The land for all three was sold to Horizon (Oldbury and Wylfa) and EDF (Bradwell) for £398m in April 2009.

Policy on radioactive waste

Nuclear power stations produce waste, which must be physically stored and eventually safely disposed of, all of which must be paid for. Government policy turned out to be much better on paying than on managing storage and disposal.

Radioactive waste management was a problem already. The new

stations would add relatively little to the problem in terms of the *volume* of waste, because they were much more efficient than the earlier stations so they produced less physical waste per unit of electricity output. They would add considerably to the problem in terms of the radioactivity, but that had relatively little impact on the costs.

The Committee on Radioactive Waste Management (CoRWM) in July 2006 concluded that waste should be stored in a deep underground repository ("geological disposal"), an approach with international support, though very few countries have yet built such a depository.

In October 2006 the government accepted the CoRWM's recommendation and gave responsibility to the NDA for securing geological disposal. The obvious difficulty was where to put the depository. After decades of secrecy and subterfuge in the nuclear industry, the government instead opted for an open and voluntarist approach: communities would be invited to discuss, without commitment, the possibility of hosting a storage site, with some benefits paid in exchange. The most likely areas open to such a deal would be those already near a nuclear site and where the additional investment, jobs and other financial benefits would be particularly welcome. Cumbria was top of the list, partly because much of the UK's nuclear waste was already there, at Sellafield.

While consultation proceeded on the site, the white paper on new nuclear emphasized the need for new nuclear investors to pay for it. Any new nuclear project would need to make a funded decommissioning plan, approved by the Secretary of State. It was essential for public confidence, and to reassure the Treasury that there would be no repeat of the bail-out of BE, that investors made credible promises to set aside funds into a separate and ring-fenced fund. That fund would pay for all long-term liabilities for waste storage and disposal and for the decommissioning of the stations after closure, allowing the land eventually to be returned to normal, safe use. It was not too difficult to set up such a fund and it was much less expensive than might be expected, because the main costs wouldn't fall due for many decades, and even on a conservative estimate of the fund's returns there was enough time for it to grow to the amount needed to pay for those costs.

Later, the physical aspect of long-term storage would prove more troublesome than the financial (Chapter 11).

Strengthening regulation

Recognizing that nuclear regulation needed to be effective on safety but also clear and predictable for investors, the government commissioned a report from Dr Tim Stone, an infrastructure finance expert who had become the Expert Chair of the Office for Nuclear Development (OND) in September 2008.

Stone's December 2008 report found that the Nuclear Installations Inspectorate (NII) was severely hampered by a lack of skilled people. He recommended organizational changes, pay increases and a much longer-term perspective (10-20 years) on NII's future needs. He also mentioned that "the current main office being in Bootle, Merseyside is not ideal for recruiting and retaining staff". In January 2009 the government accepted fully the report.

Progress on the Climate Change Act

On 23 April 2009 the European directive on renewable energy came into legal effect, requiring the UK to source 15% of total energy from renewables by 2020. The same day Energy Secretary Ed Miliband announced that all new coal stations must have CCS for at least 25% of emissions, rising to 100% by 2025. The idea was for the UK to lead a new "clean coal" industry based on the as yet unproven CCS technology; in effect it meant no new coal stations would be built for the foreseeable future. A government source told the *Guardian* newspaper that this was a "complete rewrite of UK energy policy" (*Guardian*, 2009).

The original aim of energy policy had been to leave it to the market. Climate change policy had required some intervention but the intention was still to rely on market forces where possible. However, the government was now setting goals for renewables and was restricting new build of coal. Nuclear was still spoken of as an option for private investors. But the reality was that intervention was steadily replacing market forces, for good or ill.

In May 2009 the government put the first three carbon budgets into legislation. For the periods 2008-12, 2013-17 and 2018-22 there was for the first time a legal budget for the volume of emissions (Figure 8.1).

Figure 8.1: Legislated carbon budgets (millions of tonnes of carbon dioxide equivalent)

	Budget 1	**Budget 2**	**Budget 3**
	2008-12	**2013-17**	**2018-22**
Carbon budgets (MtCO2e)	3,018	2,782	2,544
Percentage reduction below 1990 levels	22	28	34

Source: CCC (2009) p. 38

In July 2009 the government published its *Low Carbon Transition Plan*, a detailed series of actions to achieve an 18% cut in emissions by 2020, amounting to a cut of 33% from 1990, consistent with the carbon budget. The policies included a sharp rise in the obligation of electricity companies to sell renewable power and the funding of up to four demonstration CCS plants for coal power stations. There was also £3.2bn of government funds to help households increase energy efficiency, and a range of other measures to support green energy, electric cars and smart meters.

The plan also said the government would "facilitate the building of new nuclear power stations". This referred to the process, already under way, of identifying sites and to the longer business of streamlining the planning system. But these would probably not make any likely contribution to the 2020 target (HM Government, 2009).

Alongside the low-carbon transition plan the government published its renewable energy strategy. It emphasized how renewable energy would contribute to the decarbonization agenda and made a strong case for renewable energy's benefits. It noted that the UK was legally obliged to increase renewable energy to 15% of the UK's total (implying about 30% of total electricity) by 2020, under the EU renewable energy directive.

While the two targets – the UK's decarbonization plan in the Climate Change Act and the EU renewables targets – were broadly consistent, there was some tension between them. If the most economic way to decarbonize the economy was nuclear, the UK might incur unnecessary extra costs in meeting the renewables target (because nuclear is not "renewable"). The EU target and CCA were both legally binding: what if they were in conflict?

Meanwhile, the Committee on Climate Change (CCC) produced its first progress report, in October 2009. There was some good news and some bad news. The CCC noted that Britain's greenhouse gas (GHG) emissions fell at an average annual rate of 1.2% from 1990 to 2007 but by only 0.95% per year in the period 2003-07. To hit the government's interim carbon budgets up to 2022 the rate needed to be 2% a year and higher still to hit the 2050 target.

The recession that started with the financial crisis in 2008 was very likely to cut emissions temporarily. But that recession, which affected the whole of Europe, had also caused a large fall in the ETS carbon price, discouraging new investment in low-carbon energy. And the financial crisis had had a negative impact on the funding of new renewable and nuclear investment. To meet the carbon budgets a "step change" would be needed in emissions.

The main cause of the slowdown in emission cuts was that the power sector had stopped reducing its carbon intensity (ie) the amount of CO_2 emissions per kWh generated).

The fall in carbon intensity from 770 g/kWh in 1990 to 493 g/kWh in 1999 had been a result of: i) a large switch from coal generation to gas; and ii) the rise in nuclear output as BE's managers got the AGRs to work properly.

These trends had, however, run their course. Output from existing nuclear stations was expected to fall by 2020 as more Magnox stations closed. And coal had made something of a comeback against gas because it was cheaper, though from 2015 the EU Large Combustion Plant Directive would force the last major coal plants to close.

The CCC had, in its original December 2008 report, put forward three scenarios for carbon reduction: the Current Ambition, the Extended Ambition and the Stretch Ambition. Hitting the 2050 decarbonization targets required at least the Extended Ambition, involving additional changes to policy, leading to "a significant penetration of renewable heat, more ambitious energy efficiency improvement in cars and some lifestyle changes in home and transport." Initially the CCC had seen renewables (mainly wind) and nuclear as competing, and gave alternative scenarios with and without additional nuclear build. Its October 2009 report put forward a new scenario with both wind and nuclear: wind generation could be added fairly quickly, to achieve the lower carbon target by 2020; nuclear would then underpin the next stage of decarbonization, and do it more cheaply than offshore wind.

The CCC envisaged 23 GW of new wind capacity, 4 GW of other renewables and three new nuclear stations coming on stream by 2022.

That combination would cut emissions by 54% from 2008, consistent with the longer-term path to 80% cuts by 2050. For three new nuclear stations to be operating by 2022, the first application for planning permission would need to be lodged in 2010 with others coming at 18-month intervals. That in turn required the government to complete the SSA and regulations for a funded decommissioning programme by 2010, with generic design approval of the first station completed by 2011. None of these targets would be hit (see Figure 8.2).

Figure 8.2: CCC assumptions on new nuclear progress

	CCC assumption	Actual (Hinkley Point C)
Strategic Siting Assessment	2010	July 2011
Funded decommissioning programme	2010	October 2013
Generic design approval of EPR	2011	December 2012
First planning permission lodged	2010	October 2011
Planning permission granted	2011	March 2013
First construction	2012-13	2016?
First new plant operational	2018	2025?

Source: CCC (2009) author's estimates

Apart from identifying a range of government actions needed to boost wind and nuclear construction, the CCC pointed to another critical ingredient in the decarbonization strategy: changing the electricity market itself. The British electricity system, designed on the basis of minimal government intervention and a free market, was now looking decidedly unsuited for the goal of meeting the carbon emission targets. In fact it was looking like it might not even keep the lights on (see Chapter 15).

Sites confirmed

On 28 October 2009 a consortium of Scottish and Southern Energy, Iberdrola and Gaz de France paid £70m for an option on NDA land at Sellafield to build up to 3.6 GW of new nuclear capacity.

In November the list of 10 approved sites appeared in the draft National Policy Statement for Nuclear Power Generation published along with the Planning Act. The government's announcement that there would be 10 new nuclear stations was greeted with criticism.

The Conservative Shadow Secretary of State for Energy and Climate Change, Greg Clark, attacked the government plans for being "10 years too late" and saw the policy being driven by an imminent power shortage "emergency". The Liberal Democrats energy spokesman, Simon Hughes said that "A new generation of nuclear power stations will be a colossal mistake, regardless of where they are built. They are hugely expensive, dangerous and will take too long to build" (*Telegraph*, 2009).

The Department of Energy and Climate Change (DECC) asked engineering consultants Atkins to investigate whether there might be other sites than those nominated. In its November 2009 report Atkins said it found only three sites "worthy of further consideration" for deployment by 2025: Druridge Bay, Northumberland; Kingsnorth, Kent; and Owston Ferry, Lincolnshire (Atkins, 2009). Any future large-scale nuclear programme might well be constrained by a lack of sites. But for now there was plenty to be content with in Whitehall's nuclear team.

Conclusion

In 2008 it was possible to support the Climate Change Act but oppose new nuclear power. By the end of 2009 the CCC made this position increasingly untenable. The government continued to believe that if it acted quickly then the private sector would do the rest and that no subsidy was needed.

The Copenhagen Accord, signed on 18 December 2009, was an international but non-binding commitment to keeping global temperature increase to no more than 2°C. The Kyoto protocol was due to expire in 2012. The UK, at least, was going ahead with its decarbonization plans regardless.

REFERENCES

Atkins (2009). *A Consideration of Alternative Sites to Those Nominated as Part of the Government's Strategic Siting Assessment Process for New Nuclear Power Stations*. November. http://webarchive.nationalarchives.gov.uk/20110302182042/data.energynpsconsultation.decc.gov.uk/documents/atkins.pdf

BERR (2008). *Meeting the Energy Challenge: A white paper on nuclear power*. Cm 7296. January. https://www.gov.uk/government/uploads/system/

uploads/attachment_data/file/228944/7296.pdf

CCC (2009) *Meeting carbon budgets – the need for a step change. Progress report to Parliament.* 12 October www.theccc.org.uk/publication/meeting-carbon-budgets-the-need-for-a-step-change-1st-progress-report/Centrica (2009). *General meeting transcript.* 8 June. http://www.centrica.com/index.asp?pageid=752&sub=Transcript_Sam_Laidlaw

DECC (2009). *Draft National Policy Statement for Nuclear Power Generation.* (EN-6). https://www.gov.uk/government/uploads/system/uploads/attachment_data/file/228630/9780108508332.pdf

EDF Energy (2009). *Press Release: EDF Energy welcomes government announcement on nuclear sites.* 27 January. http://press.edf.com/press-releases/all-press-releases/2009/edf-energy-welcomes-government-announcement-on-nuclear-sites-42936.html

Guardian (2009). "Clean coal push marks reversal of UK energy policy". 23 April. http://www.theguardian.com/environment/2009/apr/23/clean-coal-energy-policy

HM Government (2009). *The UK Low Carbon Transition Plan.* 15 July. https://www.gov.uk/government/uploads/system/uploads/attachment_data/file/228752/9780108508394.pdf

Stone, T. (2008). *Nuclear Regulatory Review: Private advice and reasoning.* December 2008. http://webarchive.nationalarchives.gov.uk/20100512172052/http://www.decc.gov.uk/media/viewfile.ashx?filepath=what%20we%20do/uk%20energy%20supply/energy%20mix/nuclear/whitepaper08/file49848.pdf&filetype=4

Telegraph (2009). "Ten new nuclear power stations given go-ahead". 10 November. http://www.telegraph.co.uk/news/politics/6532905/Ten-new-nuclear-power-stations-given-go-ahead.html

9

New government, same policy
(2010-11)

The greenest government ever

No single party won an outright majority of seats in the general election of 6 May 2010. Six days later the Conservative and Liberal Democrat parties formed a coalition government, with David Cameron as Prime Minister. Leaders from the two parties had a long list of policies they could agree on, finding much more common ground than expected, to the chagrin of the Labour party.

On energy policy, the obvious difficulty was that the Lib Dems had historically been hostile to nuclear, reflecting the party's traditionally strong environmentalist support, which was reinforced by a scepticism about the way that large-scale nuclear entailed centralized, top-down power. The coalition overcame this by installing as Secretary of State for Energy and Climate Change Lib Dem Chris Huhne, who co-authored the Coalition Agreement, which listed the new government's policies. Tenth out of thirty-one was energy and climate change. There was a long list of specific policies that both parties could sign up to. But on nuclear the agreement, published on 20 May, included a sort of non-aggression pact:

> We will implement a process allowing the Liberal Democrats to maintain their opposition to nuclear power while permitting the Government to bring forward the National Planning Statement for ratification by Parliament so that new nuclear construction becomes possible. This process will involve: the Government completing the drafting of a national planning statement and putting it before Parliament; specific agreement that a Liberal Democrat spokesperson will speak against the Planning Statement, but that Liberal Democrat MPs will abstain; and – clarity that this will not be regarded as an issue of confidence.

> (HM Government, 2010: section 10)

In other words, individual Lib Dem MPs would be free to maintain their opposition to nuclear, which might therefore be voted down in the House of Commons, though only if Labour reversed its own support (always a possibility in the UK's adversarial system). If the government did lose the vote on nuclear, it would not be treated as a fundamental barrier to continuing with all the other policies on which the two parties agreed.

The other policies included: a carbon price floor, to compensate for the continued low carbon price in the European carbon trading system; reform of the electricity market; a ban on new coal stations, unless they had carbon capture and storage (CCS); and a "green" investment bank, to help fund the huge increase in climate change-related investment needed.

During his first visit to DECC, on 14 May 2010, the new Prime Minister told civil servants that he wanted to lead "the greenest government ever".

Meanwhile the Hinkley Point C (HPC) project was gathering momentum, and stirring up local controversy. In late 2009 the National Grid had started consulting people in Somerset about the possibility of building a 59-km (37-mile) transmission line to take power from Hinkley to a sub-station on its main transmission grid up the coast at Avonmouth. The wires would need to take a great deal of energy so they required the highest-voltage wires (capable of carrying 400,000 volts), which in turn required the tallest pylons, some 59 m (160') high. Unsurprisingly, local people weren't very happy with this, arguing for a mostly underground route instead, or for the cables to be buried under the Bristol Channel. National Grid argued that this would cost 10 times as much. The controversy, captured in the letters pages of the *Western Daily Press*, shows a lively debate about the pylons and cables, but very little opposition to the nuclear power station itself, a sign that, at least in the neighbourhood of existing nuclear power stations, new build was acceptable.

The Committee on Climate Change (CCC) produced its second progress report in June 2010, only nine months after its first report, with subsequent reports planned at 12-month intervals. The June 2010 publication stated that emissions had fallen by 1.9% in 2008 and 8.6% in 2009, but only because of the recession following the financial crisis and high fuel prices. A step change was still needed. The report urged swift progress on electricity market reform and on establishing a carbon price floor that would replace the ailing

European Emissions Trading System (ETS), where prices had collapsed following the recession. The price of carbon helped low-carbon energy producers by imposing a penalty on high-carbon producers. But the current penalty was far too low to encourage investment in low-carbon power, which therefore required more direct support from government.

The report also emphasized the need for progress on a demonstration plant that would accelerate the commercial acceptance of CCS, which was lagging behind schedule. As for nuclear, progress was on track, with a draft National Planning Statement issued, which would provide the basis for swift approval of sites and planning consent.

The 2050 Pathways report

The Department of Energy and Climate Change (DECC) had taken on a new scientific adviser in 2009, the same David MacKay whose book, *Sustainable Energy – Without the hot air*, had made such an impact earlier (see Chapter 7). It was not surprising that DECC now started producing analyses of pathways to the 2050 carbon targets, reflecting physical limits rather than wishful thinking. In July 2010 DECC published its *2050 Pathways Analysis*, which showed a range of routes to the 2050 80% carbon cuts target. Like MacKay's book it was based on physical limits, not cost optimization. The analysis showed a reference case which:

> assumes that there is little or no attempt to decarbonise, and that new technologies do not materialise. This pathway does not meet the emissions targets and would not ensure that a reliable and diverse source of energy was available to meet demand – it would leave us very vulnerable to energy security of supply shocks.
>
> (DECC, 2010: 16)

There were then six pathways using various assumptions; five included nuclear power. Pathway A was closest to the central case, assuming successful efforts to raise energy efficiency and develop renewables, nuclear and fossil stations with CCS, essentially close to the CCC's thinking, though with a greater emphasis on imported biomass in the form of liquid biofuels.

Pathway B assumed that CCS technology was not available. The result was a much larger investment in offshore wind, even more imported biofuels and a need for a great deal of back-up generation to balance the grid against wind fluctuations. That would amount to

5 GW of fossil fuel generation (probably all gas) lying inactive most of the time.

Pathway C assumed no new nuclear plant being built. Even more wind investment and back-up generation would be needed, plus solar PV equivalent to 5.4 square m (58' square) of panels per person by 2050.

Pathway D assumed only minimal new renewable capacity being built. A huge increase in nuclear would then become inevitable – a more than tenfold increase compared with 2007's production of nuclear electricity. But at least much less back-up generation was needed.

Pathway E assumed supplies of bioenergy are limited; more wind and solar filled the gap.

Pathway F assumed little behaviour change on the part of consumers and businesses. The target was then met only by a large increase in wind compared with the reference case, plus a large increase in imports of electricity, from 5.2 TWh in 2007 to 70 TWh in 2050, assuming surplus power was available from other countries.

The message from DECC's pathways seemed consistent with that of the CCC: more nuclear was unavoidable unless the country was willing to bet very heavily on offshore wind; and if CCS doesn't work or isn't cost effective then nuclear will be even more important.

The CCC urges new contracts to hit the next carbon budget

At the end of 2010 the CCC published the fourth carbon target, for the period 2023-27. The budget would need the addition of 30-40 GW of low-carbon electricity capacity, which would reduce average emissions from around 500 g/CO_2/kWh to around 50 g CO_2/kWh by 2030. But for this to happen, the proposed reforms of the electricity system needed to be radical:

> Existing electricity market arrangements are not well designed to ensure this progress occurs in a cost-effective fashion. Therefore we recommend that new arrangements are introduced entailing competitive tendering of long-term contracts for investment in low-carbon capacity.
>
> (CCC, 2010b: 13).

It later described the need to introduce long-term contracts as an "urgent priority" (page 34). What lay behind this?

Civil servants and politicians originally hoped that the electricity market, which had in many ways worked very well since privatization, would be kept intact but with a carbon price to encourage investment in low-carbon generation. That price signal, added to the market price of electricity, should have automatically incentivized private investment with a minimum of government intervention, preserving the benefits of the market: the choice of the lowest-cost solutions and no need for poorly informed government planners to make investment decisions. It was increasingly obvious that this happy, minimal intervention picture was not going to deliver what the government needed: renewables investment to hit the EU 2020 target plus nuclear to hit the fourth and later carbon budgets.

The first problem was that the carbon price set by the European ETS had collapsed because the European recession following the global financial crisis of 2007-09 had cut electricity demand and with it the demand for carbon emission permits. The ETS was giving far too little incentive for low-carbon investment. The CCC argued that the government should remedy the failings of the ETS by introducing a floor under the carbon price, for the UK only, and maintain it for as long as an incentive was necessary.

Second, even if the carbon price had been higher, expecting investors in long-term, capital-intensive projects to commit their funds on the basis of a volatile carbon price was asking rather too much. That was a problem for 20-year investments in wind farms; it was a complete deal-breaker for a 60-year investment in a nuclear power station, with up to 10 years of construction before that. Before the financial crisis EDF and other private companies had been perhaps too optimistic about their ability to finance new reactors on their own balance sheets, without asking for funds from outside investors and banks. Now their share prices had fallen, because of that financial crisis and, in the case of EDF, owing to concerns about how French electricity reform might hurt profits in its home market. In addition, the banks were now badly damaged and in no mood to lend to long-term, risky projects.

The only way to attract the huge investment sums needed to build 30-40 GW of new capacity was to offer long-term power contracts at fixed prices, taking away some of the investors' risk, though not the construction risk. That sounded a lot like subsidy, though the price would be paid by the electricity consumer, not by the government. The subsidy could be justified by the need to meet policy targets on renewables and decarbonization. But it was still a subsidy.

The government was setting targets for renewables and for low-carbon electricity investment. It was being asked to offer long-term price contracts. What was left of the market in all of this? The CCC (and the government) wanted to keep as many decisions as possible in private hands, so it argued for these contracts to be offered in an auction. The government would say that a certain amount of low-carbon capacity was needed by a particular date and would invite private investors to tender for the lowest price promise at which they would be willing to invest. That should ensure that the cheapest projects were funded and that prices didn't rise too far above the minimum needed to pay a fair return.

In 2015 this auction framework was eventually made to work for renewables. Renewable investment can be delivered in relatively small increments, though offshore wind is increasingly large scale. But that framework was never realistic for nuclear, which comes in extremely large chunks. Although it would be theoretically possible for the government to auction a site for a new nuclear power station, that would require at least two developers with fully approved reactor designs, who are able to compete with each other. That was out of the question. Each nuclear developer had a separate site and a separate reactor type, approval of which was likely at different dates. It might eventually be possible for an auction of long-term prices for nuclear, but for the next few years each contract would be one negotiated between the government and investors, which, as we shall see, took a long time and produced controversial results.

The CCC must have had its doubts about all of this but it kept quiet. It reported that DECC's indicative timeline showed new nuclear starting operation in 2018 (CCC, 2010b: Fig. 6.12). This was one year later than originally hoped.

The 2050 Pathways analysis and the CCC's work emphasized how essential new nuclear investment was to both the medium-term (2023-27) and longer-term (to 2050) targets for decarbonizing the UK economy. But there now seemed to be real momentum behind government policy. In October 2010 the government produced its National Infrastructure Plan, a document that critics said amounted to little more than a wish list of projects. The government was trying to push infrastructure up the policy agenda, though the effect of its central policy of cutting the budget deficit was to lead to lower capital spending by the government itself. Figure 9.1 shows that public sector net investment declined sharply from the first year of the coalition government.

The fact that the government was cutting its own investment increased the need for private funding. At the launch of what was

Figure 9.1: Public sector net investment, 2010-14 (£bn). Figures above the bars show the percentage of GDP that the respective investment represents

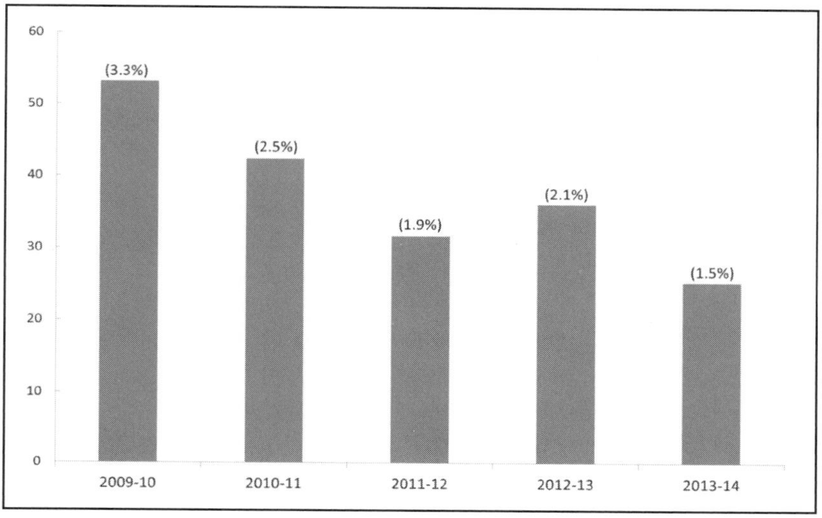

Source: House of Commons Library, 2014: 9

the UK's first ever infrastructure plan, Prime Minister David Cameron spoke of "how we will unlock some £200 billion worth of public and private sector investment over the next five years to deliver it" (HM Treasury, 2010). Once again the language implied that there was plenty of appetite from private sector investors and the government needed only to channel it. The former Labour government had announced a new unit in the Treasury, Infrastructure UK (IUK), which included some staff with private finance experience who would coordinate this channelling. IUK actually started in June 2010 under the coalition government. The National Infrastructure Plan listed projects that the government saw as contributing to economic growth and productivity and as vital to the economy. But actually making them happen depended on providing the right incentives.

Perhaps the most tangible evidence that a new nuclear power station might actually be coming was the report in November 2010 that EDF had moved a badger colony from the Hinkley Point site. Anyone who knows the UK well will understand the significance of this – when wild animals, especially cute ones like badgers, become involved, things are getting serious. EDF told anxious wildlife lovers that it had moved the animals with a licence from the wildlife "watchdog" Natural England. The badgers had experienced no distress and would be happy in their new home.

In December the government showed it was fully on board with the CCC's agenda when it launched consultations on electricity market reform (EMR) and on a proposed carbon price floor.

EMR – electricity market reform

The EMR consultation document began by acknowledging that the existing market had worked well, delivering some of the lowest power prices in Europe. But the market now faced challenges: replacing about 25% of existing generation capacity, much of which was old or dirty; a sharp rise in renewable electricity from 7% to 30% by 2020; and the need to virtually decarbonize the power sector by 2030 in line with the CCC target. These challenges would require reform.

The solution was four modifications to the market:

i)   Carbon price support: the government would legislate to provide minimum carbon prices, rising over time;
ii) Long-term power price contracts to provide stable returns for investors;
iii) Capacity payments: targeted payments to encourage security of supply through demand reduction measures (so-called "negawatts") or the construction of flexible reserve plants to ensure the lights stay on, particularly as the system became more dependent on intermittent wind and solar power that could not be relied on to generate at any particular moment;
iv) An Emissions Performance Standard to limit how much carbon the most carbon-intensive power stations – coal-fuelled stations– can emit; this would reinforce the existing requirement that no new coal stations be built without CCS (which was not yet commercially available).

These amounted to a radical change in the electricity market. Pinning down the details of the long-term price contracts was to prove especially troublesome, though they would ultimately work – investors are very attracted to a long-term price promise, so long as they're confident it will be honoured.

It was increasingly difficult to argue that the UK would still have an energy market. The difficulty was in reconciling two very different ways of organizing economic activity. In a genuine market, supply and demand are matched through a price mechanism, which can work well if there are no externalities (consequences for other people that are not captured by the price – pollution being the most obvious example).

The polar opposite of a market is central planning: the government or state-owned companies decide on what to build and how to price it. The UK experience of this in the energy industries was far from encouraging. The Central Electricity Generating Board (CEGB) overinvested and managed its power stations very inefficiently.

Privatization and the eventual deregulation of the electricity industry had led to a big shakeout in costs, with much lower staff levels at power stations, and a more thoughtful approach to investment rather than building lots of capacity just in case. The CEGB always had enough funds (from captive electricity customers) to build whatever it saw fit; private generators would be more careful, as they were risking their and their shareholders' money, and if the capacity wasn't needed they would make big losses.

A slight simplification would be to say that the private, deregulated power market did a good job in that it squeezed costs out of the system, and it used the existing assets well. This is known among economists as "static efficiency". However, the challenge the deregulated market hadn't yet had to face was the replacement of those assets with new capacity; what is called "dynamic efficiency". The experience of other capital-intensive industries, such as steel, bulk chemicals and mining, suggests that the private market doesn't do such a good job with dynamic efficiency. These industries are prone to investment peaks and troughs that can cause large price variations. A shortage of capacity leads to rising prices, and eventually, when supply is really at its limit, companies all invest in new capacity, expecting high profits. But when all this new capacity comes on stream at once, often coinciding with an economic downturn that has cut demand, the price collapses and everybody makes losses for a while.

This is not such a problem for industries producing physical goods that they mainly sell to other businesses: their product is storable, giving some scope for bridging gaps between demand and supply. Electricity is different. High and volatile prices affect ordinary households, so it becomes a political issue. And electricity is not storable in large quantities, so if capacity is scarce, the only way to meet demand is by sharply raising prices; if that doesn't work, the next step is rationing – which means blackouts.

The UK electricity market might well have eventually delivered all the new capacity needed through the incentive of high power prices. But those prices might attract political criticism (as was to happen a few years later), leading to political promises to curb them, thus removing the economic incentive to invest. No government wanted to risk temporary "rationing" arising from a mismatch between demand and supply. It was one thing to have a temporary shortage of steel (which can always be imported); it's quite another to explain to the public why the lights had to go out – something associated with poorly managed emerging economies, not a mature country like the UK.

So political risk aversion reinforced genuine doubts about how well the market would meet the capacity challenge (the energy regulator, Ofgem, estimated the new generation investment at £110bn). The result was EMR: a radical break from previous power market policy and a new and untried set of rules for investors to navigate.

In February 2011 the government announced that the former Nuclear Installations Inspectorate (NII) would be re-created as a stronger and more independent Office for Nuclear Regulation (ONR), separate from the Health and Safety Executive (HSE), of which it had previously been a part, with better staff resourcing following the implementation of the Stone reforms (Chapter 8). The ONR would also for the first time consolidate in one place civil nuclear and radioactive transport safety and security regulation.

On 8 March 2011 DECC published the government's *Carbon Plan*. The introduction, signed by the Prime Minister, Deputy Prime Minister and Secretary of State for Energy and Climate Change, included a commitment to "new unsubsidised nuclear power". It set out "a vision of a changed Britain, powered by cleaner energy used more efficiently in our homes and businesses, with more secure energy supplies and more stable energy prices, and benefiting from the jobs and growth that a low carbon economy will bring". It also provided the increasingly lengthy list of government actions that were needed to make all of these wonderful things actually happen (HM Government, 2011).

One of those actions was to get the carbon price back to a point where it could encourage low-carbon investment. In the budget of 23 March 2011 the Chancellor announced the floor would start in April 2013 at about £25/t and rise steadily to £30/t in 2020 and £70/t by 2030. Not incidentally, given the coalition government's main priority of cutting the government budget deficit, it would raise around £1bn of annual revenue too. But it was well received. Vincent de Rivaz of EDF Energy noted that "For nuclear, helping to restore the carbon price to what was originally intended is important to encourage investment in existing plants and in new build" (*World Nuclear News*, 2011).

Conclusion

The new coalition government had carried out the policy of its predecessor with remarkable continuity, just the sort of long-term thinking that investors want. It was listening to the CCC and stepping up progress on reforming the electricity market. The early works for HPC were going ahead. It seemed everything was proceeding for unsubsidized nuclear to happen on schedule. What could go wrong?

REFERENCES

CCC (2010a). *Meeting Carbon Budgets – Ensuring a low-carbon recovery (2nd progress report)*. 30 June. http://www.theccc.org.uk/publication/meeting-carbon-budgets-ensuring-a-low-carbon-recovery-2nd-progress-report

CCC (2010b). *The Fourth Carbon Budget*. 7 December. http://archive.theccc.org.uk/aws2/4th%20Budget/CCC-4th-Budget-Book_interactive_singles.pdf

DECC (2010). *2050 Pathways Analysis*. July. https://www.gov.uk/government/uploads/system/uploads/attachment_data/file/42562/216-2050-pathways-analysis-report.pdf

HM Government (2010). *The Coalition: Our programme for government*. https://www.gov.uk/government/uploads/system/uploads/attachment_data/file/78977/coalition_programme_for_government.pdf

HM Government (2011). *Carbon Plan*. 8 March. https://www.gov.uk/government/uploads/system/uploads/attachment_data/file/47621/1358-the-carbon-plan.pdf

HM Treasury (2010). *Government Launches National Infrastructure Plan*. 25 October. http://webarchive.nationalarchives.gov.uk/20130129110402/http://www.hm-treasury.gov.uk/press_56_10.htm

House of Commons Library (2014). *Infrastructure Policy*. Standard Note: SN/EP/6594. Last updated: 9 December. www.parliament.uk/briefing-papers/sn06594.pdf

World Nuclear News (2011). "UK government introduces carbon floor price". 23 March. http://www.world-nuclear-news.org/IT-UK_government_intro duces_carbon_tax-2303114.html

10 / Fukushima
(2011)

The Fukushima disaster

At 14.46 on 11 March 2011, a particularly large earthquake started in the Pacific Ocean about 72 km (45 miles) from the Japanese coast. The earthquake lasted six minutes and was recorded as level nine on the international scale; it was the largest earthquake ever to hit Japan and the fourth largest in the world since modern recording began in 1900. The earthquake displaced sea water, resulting in a tsunami with a maximum height of 15 m (49'), which struck the coast of Japan about 50 minutes later. The combination of earthquake and tsunami caused more than 15,000 deaths. The World Bank estimated the economic cost of the disaster at $235bn, the most expensive in history.

At the Fukushima Daiichi nuclear power plant on Japan's east coast, three of the six boiling-water reactors (BWRs) were out of operation at the time of the earthquake. The other three were automatically shut down when the earthquake struck.

A nuclear reactor continues to produce heat even when its control rods are fully inserted in order to stop the nuclear fission. This heat must be removed or the reactor core will overheat. At Fukushima Daiichi, if power was lost the reactors relied on back-up diesel generators to keep sea water pumping through the reactors and take the heat safely out to sea.

Although the earthquake did little damage, the tsunami, around 14 m (45') high, overwhelmed the station's protective sea wall, which was only 10 m (33') tall. A second tsunami arrived a few minutes later. The sudden flooding knocked out the sea-water pumps and then the diesel generators that provided power to them. Critical electrical equipment was flooded and the roads around the station were badly damaged. Two people at the plant were trapped by the flooding and died.

At about 17.00 on 11 March a nuclear emergency was declared when it became clear that the reactors would overheat and that radiation emissions were likely. An area with a radius of 2 km (1¼ miles) around the plant was evacuated; this was rapidly extended to 3 km (1¾ miles), then 10 km (6¾ miles) and then 20 km (12½ miles).

About an hour after the fission reaction had been shut down, the reactors were producing about 1.5% of their normal heat. While this is a small percentage, it represented some 22 MW of energy in reactor 1 and 33 MW in each of reactors 2 and 3 (20 MW is the power used by roughly 4,000 UK homes). The heat turned the water surrounding the fuel rods to steam, and a chemical reaction between the fuel cladding and the steam subsequently produced hydrogen, along with yet more heat. The mixture of hydrogen and steam was released through safety valves into the outer section of the reactor containment vessel, steadily increasing the pressure in this area.

The hydrogen in reactor 1 exploded on 12 March, blowing off a section of the building's roof. Venting the gas from reactor 2 avoided a similar explosion. An even larger hydrogen explosion occurred at reactor 3, causing further damage. There was also an explosion at reactor 4, which was not operating and didn't contain any fuel; this incident was probably caused by a flow of gas from reactor 3.

The venting and explosions released radioactivity into the atmosphere. Most of the fuel, though badly damaged by fire, remained in the reactors or in the concrete containment area beneath them. Radioactive emissions continued until December when they returned to minimal levels and the plant was put into "cold shutdown".

Deaths from nuclear incidents in perspective

Reports from the World Health Organization (WHO) and the United Nations Scientific Committee on the Effects of Atomic Radiation (UNSCEAR) suggest that the Fukushima nuclear disaster, the worst since Chernobyl in 1986, definitely killed two people and that the radioactivity released may lead to an unknown number of additional cancers, some of which may cause premature deaths. Both reports found, however, that most people were subject to lower radiation doses than they receive from naturally occurring "background" radiation, so it will be hard to tell if the disaster has caused any additional cancer cases, as these are likely to be low compared with "baseline" rates.

WHO found that "For all other locations in Japan [than Fukushima] and around the world, the radiation-related cancer risks were estimated to be much lower than the usual fluctuation in the baseline cancer risks" (WHO (2013) section 8.1). This is *not* the same as saying that no additional cancer risk applies outside the most affected areas; only that the risk is very small indeed and impossible to quantify with any useful precision. WHO concluded that for the general public,

> The present results suggest that the increases in the incidence of human disease attributable to the additional radiation exposure from the Fukushima Daiichi NPP accident are likely to remain below detectable levels.
>
> (WHO (2013) section 8.1)

UNSCEAR found that the additional exposures most Japanese people experienced in the first year after the disaster and in subsequent years "are less than the doses received from natural background radiation". There may be a "theoretical risk" of additional thyroid cancers in future, but it is impossible to predict how many the disaster may cause (UNSCEAR, 2013).

The UNSCEAR report and its conclusions were criticized by physicians from 19 affiliates of the International Physicians for the Prevention of Nuclear War (IPPNW, which was awarded the Nobel Peace Prize in 1995) in a report published in June 2014 (IPPNW, 2014).

IPPNW argued there would probably be many additional cancers, estimating between 4,300 and 16,800 extra cases, resulting in 2,400-9,100 extra deaths in the Japanese population compared with what could have been expected without the Fukushima accident. IPPNW calculates its estimates by extrapolating from very small doses received by a large number of people, a method criticized by the International Commission on Radiological Protection (International Commission on Radiological Protection, 2007).

The probable number of extra deaths therefore varies hugely, depending on which estimate is more accurate – the IPPNW's or UNSCEAR's. Figure 10.1 puts Fukushima in the context of immediate deaths from the worst nuclear accidents in the past.

In all of these cases there is also a possibility of subsequent deaths. For Chernobyl the WHO estimates of total lifetime cancers caused was around 4,000. Figures such as 4,000 or even 9,100 extra deaths are shocking, even when they are spread over many decades. Is nuclear energy therefore especially dangerous? Regrettably all sources of energy are hazardous. Figure 10.2 shows the estimated total deaths caused by accidents in energy production.

Figure 10.1: Immediate deaths from worst civil nuclear incidents

Location	Country	INES (*)	Year	Immediate deaths
Chernobyl	USSR	7	1986	56
Fukushima	Japan	7	2011	2
Kyshtym	USSR	6	1957	?
Three Mile Island	USA	5	1979	0
Windscale fire	UK	5	1957	0
Idaho Falls	USA	4	1961	3
Tokaimura	Japan	4	1999	2
Jaslovské Bohunice	Czechoslavakia	4	1976	2
Mihama	Japan	1	2004	4

(*) International Nuclear Event Scale – an indicator of the severity of the incident.
Source: Sovacool, 2010; and UNSCEAR, 2013

Figure 10.2: Deaths caused by energy accidents 1907-2007

Energy type	Deaths
Hydroelectric	171,216
Nuclear	4,067
Oil	3,330
Coal	2,834
Natural gas	709

Source: Sovacool (2008)

While the huge number of deaths caused by hydroelectricity may initially seem surprising, it arises, in fact, almost entirely from one incident. On 8 August 1975 the Shimantan dam in China collapsed with the loss of an estimated 171,000 lives. Similarly, the great majority of the deaths caused by nuclear power production is attributable to Chernobyl (4,056 of the 4,067 total; note that this data excludes Fukushima). Even in oil, a few major accidents dominate the total, with a gasoline pipeline explosion in southern Nigeria in 1998 causing 1,088 deaths out of a total for the oil sector of 3,330.

These estimates concern accidents, whereas the majority of deaths from coal, for example, arise from atmospheric pollution. In 2012 WHO raised its estimate of global annual deaths from air pollution, much of it caused by coal burning, to seven million. Around 3.7 million of these were caused by outside pollution, the remainder by indoor pollution linked to the use of fossil fuels and wood for cooking and heating (WHO, 2014). In addition, the above accident and atmospheric pollution figures do not take any account of coal's leading role in carbon emissions, which cause climate change and possibly further deaths.

The UK reaction: the Weightman review

The Fukushima accident came as a shock to both the British public and its nuclear policymakers. Memories of the last major nuclear accident – Chernobyl in 1986 – had started to fade. Although this was a far larger disaster and was geographically much closer to the UK, it had been seen as the result of the Soviet Union's inferior technical standards and a disregard for risk to the population – an attitude possible only under an authoritarian regime.

Japan, however, was a rich, modern economy with the highest standards of industrial efficiency, albeit one with a systemic earthquake problem. If a nuclear disaster could happen there with, presumably, advanced safety systems and smoothly operating nuclear processes, then perhaps it could happen anywhere.

As was later revealed, Japan's outstanding industrial record did not apply to the nuclear industry, which a Japanese independent inquiry described as secretive and "immune to scrutiny by civil society" (CNN, 2012). It had cut itself off from global efforts to improve nuclear performance, safety and public confidence since Chernobyl.

Consequently, there was an immediate question for the UK: was the country now making a mistake in pursuing new nuclear stations?

On 14 March 2011, three days after the Fukushima accident, Secretary of State for Energy and Climate Change, Chris Huhne, asked the Chief Nuclear Inspector, Professor Mike Weightman, to produce a report on the implications of events in Japan for current and future nuclear plans in the UK. There would be an interim report by May and a full report within six months.

Weightman had been the head of the Nuclear Directorate within the Health and Safety Executive (HSE) and became the Director of the Office for Nuclear Regulation (ONR) when it was created on 1 April 2011. He had an impressive range of academic and professional qualifications in materials science, engineering and physics. He had worked in the nuclear industry for 13 years before joining the ONR's predecessor organization, the Nuclear Installations Inspectorate (NII). He had also led an independent inquiry into the rail crash at Potters Bar in 2002.

Weightman's interim report, published on 18 May 2011, looked at "the evidence and facts, as far as they are known at this time, to establish technically based issues that relate to possible improvements in safety and regulation in the UK. It also indicates some lessons for international arrangements for such systems" (ONR, 2011). The focus of the interim report was on nuclear power stations. The report summarized the UK's approach to nuclear safety regulation:

> The UK nuclear regulatory system is largely non-prescriptive. This means that the industry must demonstrate to the regulator that it fully understands the hazards associated with its operations and knows how to control them. The regulator challenges their designs and operations for safety to make sure that their safety provisions are robust and that they minimize any residual risks. So, we expect the industry to take the prime responsibility for learning lessons, rather than relying on the regulator to tell it what to do. What we have done in this report is point out areas for review where lessons may be learnt to further improve safety. But it is for industry to take ultimate responsibility for the safety of their designs and operations."
>
> (ONR (2011) "Executive summary")

Weightman further pointed out that all but one of the UK's reactors were gas-cooled and therefore of a fundamentally different design from the BWRs at Fukushima. As they don't use water as a coolant, they are not at risk of generating hydrogen in an accident, the main cause of the explosion at Fukushima. In addition, while Sizewell B was a pressurized-water reactor, it was of a much more recent design than the Fukushima BWRs and had more advanced safety features.

The main conclusion of the report and the crucial one for the UK's nuclear new build policy was that "we see no reason for curtailing the operation of nuclear power plants or other nuclear facilities in the UK" (ONR (2011) "Conclusion 1"). In large part this was because the UK, being some 1,000 km (620 miles) from the nearest edge of a tectonic plate, doesn't have the earthquake and tsunami risk of Japan. Also, the UK nuclear industry had a good safety record and had "reacted responsibly and appropriately displaying leadership for safety and a strong safety culture in its response to date" (ONR (2011) "Conclusion 2").

The report further commended the creation of the ONR for helping to reinforce confidence in the UK safety system. It found no gaps in the scope or depth of the safety assessment principles for nuclear facilities in the UK, nor any evidence from Japan of a need to change the licensing process in the UK or to change the siting strategies for new nuclear power stations. It confirmed that, compared to water-cooled reactors, the UK's gas-cooled reactors were at lower risk of overheating in the event of a loss of coolant, a well-known feature of their design.

The report considered the implications for the two new reactor designs under consideration by the ONR for future build in the UK: the French EPR (European pressurized reactor or Evolutionary Power Reactor) and Japanese–American AP1000. Both are more advanced models than either the BWRs used at Fukushima or the PWR at Sizewell B. Significantly, both have mechanisms to prevent the build-up of hydrogen gas that caused the explosions at Fukushima.

The report added a long list of recommendations that overall amounted to reviews of all the existing nuclear arrangements but contained no presumption that any of them would be found wanting. The UK's emphasis on continuous improvement means that such reviews are a normal part of the process in any case. Continuous improvement, it is important to note, is a key feature of the global procedures adopted by all nuclear operators through their membership of the World Association of Nuclear Operators (WANO), set up after Chernobyl.[1] It was widely known in the industry that one country that showed reluctance to follow WANO procedures fully was Japan.

Weightman's final report, published in September 2011, didn't change any of these findings. Taking into account all of the UK's nuclear facilities, not just power stations, it concluded that the UK's safety processes were in general sound, including the procedure for identifying the design basis for safe operation. In other words, Weightman found there was no reason to halt the existing process for approving the new reactors under consideration for new build.

---

[1] WANO's mission statement is "To maximise the safety and reliability of nuclear power plants worldwide by working together to assess, benchmark and improve performance through mutual support, exchange of information and emulation of best practice". http://www.wano.info/en-gb/aboutus/ourmission

Figure 10.3: UK and Japanese public attitudes to new nuclear build
(Agreement with the statement "I am willing to accept the building of new nuclear
power stations if it would help to tackle climate change" (% tend to/strongly agree))

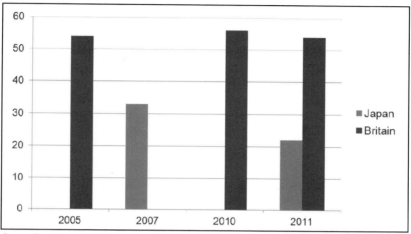

Source: Poortinga, W., M.Aoyagi and N. Pidgeon (2013)

Welcoming the interim report, Secretary of State Chris Huhne told the House of Commons that he was pleased with the findings. The International Atomic Energy Agency (IAEA), he said, had recently "noted that the UK has a mature, transparent and independent regulatory system, an advanced review process, and highly trained and experienced nuclear inspectors." He went on to say that he saw "no reason why we should not proceed with our current policy: namely that nuclear should be part of the future energy mix in the future as it is today, providing that there is no public subsidy" (Huhne, 2011).

Despite the worst nuclear accident in 25 years, nuclear new build remained on track in the UK. The Weightman report and later studies of the Fukushima accident increasingly reinforced the sense that the Japanese nuclear industry had been incompetent and secretive, and that these conclusions didn't apply to the UK. There were of course those who distrusted the Weightman report because its author was a nuclear "insider". There were others who thought the government should wait much longer – years, even decades – to make sure there were no longer-term lessons from Fukushima.

British public opinion, however, appeared only temporarily disturbed by the Fukushima disaster and even then only slightly. Research undertaken by two academics at Cardiff University, using

Ipsos MORI opinion poll data, showed the dramatic difference in attitudes to new nuclear build between the UK and Japanese public (Figure 10.3).

So, based on these figures, Huhne was not risking a major public backlash in continuing with the new nuclear strategy. It would be the reaction of another country, Germany, that would be important for the UK nuclear programme.

Germany's reaction to Fukushima affects the UK

In Germany, the impact on nuclear policy of the Japanese disaster was more significant than it was in the UK, and its response posed a risk for UK new nuclear build.

Historically, Germany has combined an excellent safety and operating record in civil nuclear power with a visceral hostility to nuclear from a large section of the German population. In 2010 about a fifth of the country's electricity was generated from nuclear, from standard PWRs built from US designs. The effect of the Chernobyl disaster in 1986 was far greater in Germany than in the UK, partly owing to the country's closer proximity to the USSR, but also because of a long-standing fear of nuclear radiation arising from Germany's front-line role in the Cold War. Much of Germany was contaminated with radiation from Chernobyl. With scenes of radioactive crops being destroyed and with the frequent anti-nuclear protests often turning violent, the Green anti-nuclear movement saw a surge in support.

In 2000, when the Green Party entered a coalition government led by Social Democrat Chancellor Gerhard Schroeder, nuclear was top of its list of concerns. In 2002 the government duly passed the "Act on the structured phase-out of the utilization of nuclear energy for the commercial generation of electricity". All nuclear power was to be phased out in Germany by 2022, with two stations closing in 2003 and 2005. This was something of a compromise compared with the Green's desire to close nuclear much more quickly.

Angela Merkel, when she became Chancellor in 2010, leading a conservative government, announced that the phase-out would be delayed until 2036, provoking several major demonstrations. She argued that the nuclear plants were still relatively new, working satisfactorily and would produce economic and low-carbon power for years to come. However, the country's strong commitment to renewable energy remained in force.

Three days after Fukushima, however, the same Angela Merkel, facing five important state elections in the following two months, announced a three-month provisional closure of seven older stations, all constructed before 1980, with a total of 8.4 GW of capacity. This was pending a decision on longer-term German nuclear policy following the Fukushima disaster.

The Green party won huge majorities over its Conservative rivals in Baden-Württemberg and Bremen. The famously pragmatic Merkel announced on 30 May 2010 that all of Germany's nuclear stations would once again be phased out by 2022. The new law was passed with an overwhelming majority in the Bundestag (the lower house of parliament).

The plan was strongly criticized by German companies involved in nuclear power and engineering. They argued the country would be unable to replace the power from nuclear so quickly, would have to import electricity (mostly from France, where almost all the power is nuclear) and would risk being unable to cut carbon emissions. All of this later turned out to be accurate, but the popularity of nuclear power had never been so low.

German anti-nuclear policy mattered to the UK because one of the consortia that had put forward credible plans to build new nuclear stations in the UK was backed by two German electricity companies. Horizon Nuclear Power was set up in 2009 by RWE and E.ON, rivals at home but collaborating on the opportunity to enter the new nuclear market in the UK, where each controlled a major subsidiary in the conventional power generation and distribution business. Horizon planned to build two large nuclear stations totalling 6,000 MW at the sites of existing nuclear stations at Wylfa and Oldbury.

For years, the UK had sought to get some kind of competition going between EDF and other nuclear operators. It had succeeded with Horizon and with NuGen. NuGen was set up in early 2009, originally as a joint venture between Iberdrola, GDF Suez, and Scottish and Southern Energy. Scottish and Southern sold its 25% stake to the other two partners in 2011, leaving Iberdrola and GDF Suez as equal partners. NuGen unsuccessfully bid for a number of sites at existing nuclear power stations before buying land at Sellafield, where it planned to build 3,600 MW of capacity using the AP1000 reactor. (In 2013 Iberdrola sold its 50% interest in NuGen to the other major Japanese nuclear company, Toshiba, which owns the AP1000 reactor technology. Toshiba purchased a further 10% stake to become the majority owner.)

However, the German retreat from nuclear now threatened to reduce the competition in the UK. Without a home base for their nuclear expertise and facing the potentially serious financial cost of closing their German stations early (as well as increasing competition from subsidized solar and wind power), the two German utilities had no further strategic reason to continue with their UK plans. In March 2012 RWE and E.ON put Horizon up for sale.

RWE and E.ON had at first been resigned to getting a low price for Horizon and to writing off the investments they had already made, mainly in the purchase of land and in hiring a team of people. It was not at all clear that anybody would want to buy a company that had little more than some land and encouraging noises from the British government. To their delight, however, and to the relief of the civil servants in the ONR, Horizon became a hot property. There was considerable interest from five countries: China, France, Russia, the USA and, more surprisingly, Japan.

The interested parties were as follows:

i)   The French nuclear technology company, Areva, already involved in Hinkley Point C;
ii) Toshiba, the Japanese company that owned Westinghouse and its AP1000 reactor design;
iii) China Guangdong Nuclear, one of the two major Chinese nuclear operators that would later join the Hinkley project;
iv) State Nuclear Power Technology Corporation (SNPTC), which was a Chinese state-owned company that was the general contractor for the AP1000 stations being built in China and was also the developer of a larger Chinese version called the CAP1400 reactor, which China hoped eventually to sell to the rest of the world;
v)   Chicago-based utility, Exelon, one of the largest nuclear operators in the USA;
vi) the Russian nuclear company, Rosatom, which wanted to use its WER-1200 reactor, a passive-safety and larger version of the earlier Russian 1,000 MW design, which is already under construction in Russia.

In June 2012 these companies (except Rosatom, which dropped out) formed themselves into two consortia interested in buying Horizon. One was a combination of Westinghouse (Toshiba), SNPTC and Exelon (which later withdrew); and the other consisted of Areva and CGN.

Most of the interested parties had a nuclear reactor to sell. The UK had become one of the few countries in the world where there was a realistic prospect of selling new nuclear power stations. Other countries with nuclear potential, such as China, were closed and controlled markets. So the reactor vendors, such as Areva, Toshiba, SNPTC, and even Rosatom, were keen to become part of a project in which they could lobby for their reactor to be chosen. When it was made clear that approval for the Russian reactor would be some way off, if it was ever given, Rosatom offered to build "Western" reactors first, as a bridge to eventual acceptance of the WER-1200. Against the backdrop of rocky Anglo-Russian relations and lack of confidence in the Russian nuclear regulatory system, however, Rosatom was never very likely to be accepted.

One nuclear vendor was missing from the list: the electrical conglomerate Hitachi. Hitachi's nuclear business was in the same position as Toshiba's: its home market, Japan, had just experienced a huge blow to confidence in nuclear power and it was unlikely any new nuclear stations would be built there soon. The UK was the only market available to keep investment going and to show off its best technology when, perhaps, other countries decided to opt for more nuclear. Hitachi and Toshiba were intense, some would say bitter, rivals. Horizon, helped along by the British government, inflamed this rivalry quite deliberately. Toshiba started out as the leader but it was Hitachi that eventually offered the highest price and secured Horizon for £696 million.

Hitachi had more at stake than Toshiba, whose AP1000 reactor was already in the race and had started the long process of generic design acceptance by the ONR. Hitachi's offering in the new nuclear stakes was the advanced boiling-water reactor (ABWR). A joint venture with the US giant GE, this was a third-generation development of the earlier boiling-water reactors (BWRs) built in the USA and Japan. A variant of the light-water reactor technology, which included the standard PWR, BWRs were new to the UK. There was no intrinsic reason why BWRs shouldn't be built in the UK but it would help a lot if Hitachi was in the driving seat of a project to build a new nuclear station, able to ensure that its reactor was the one to be built. If Horizon eventually gets planning permission and the ABWR gets design approval (the process started in 2013), Hitachi will build four 1.4 GW reactors: two at Wylfa and two at Oldbury. Together with the ABWRs under construction in the USA and Taiwan, that would put Hitachi firmly back in the race with Toshiba's AP1000.

Toshiba, while disappointed, could take comfort from the fact that the AP1000 was emerging as everyone else's favourite third-generation reactor, already likely to be chosen by the Moorside nuclear consortium and with active support in China. The two Japanese giants, like the Germans, lacked a home market, but were determined to find an international one. The difference was that the German companies are utilities that make money from building and operating stations, of any acceptable type; they have no intrinsic interest in nuclear. The Japanese duo are engineering companies that badly want to sell reactors for power stations. With the USA now unlikely to build any new nuclear power stations for some time (beyond the four currently under construction) the UK market is their priority.

Conclusion

Despite the Fukushima disaster the British nuclear programme continued with only a brief loss of momentum. Perhaps the new, open style of nuclear consultation had reassured people. The loss of the German utilities could have been a serious blow, but the sale of Horizon showed just how attractive the British market appeared to the global nuclear reactor industry.

REFERENCES

Carnegie Endowment (2013). *Why Fukushima Was Preventable.* http:// carnegieendowment.org/2012/03/06/why-fukushima-was-preventable/ a0i7

CNN (2012). "Japanese parliament report: Fukushima nuclear crisis was 'man-made'. http://edition.cnn.com/2012/07/05/world/asia/japan-fukushima-report/index.html

Huhne, C. (2011). "Statement to House of Commons", 18 May. https:// www.gov.uk/government/news/dr-mike-weightmans-interim-report-japanese-earthquake-and-tsunami-implications-for-the-uk-nuclear-industry-written-ministerial-statement-by-the-rt-hon-chris-huhne-mp-18-may-2011

International Commission on Radiological Protection (2007). *The 2007 Recommendations of the International Commission on Radiological Protection.* http://www.icrp.org/docs/ICRP_Publication_103-Annals_of_ the_ICRP_37(2-4)-Free_extract.pdf

IPPNW (2014). *Critical Analysis of the UNSCEAR Report*. 16 June. http://www.psr.org/assets/pdfs/2014-unscear-full-critique.pdf

IRSN (Institut de Radioprotection et de Sûreté Nucléaire) (2011). *Synthèse des informations disponibles sur la contamination radioactive de l'environnement terrestre japonais provoquée par l'accident de Fukushima Daiichi*. 13 July. Paris.

McBride, J. P., Moore, R. E., Witherspoon, J. P. and Blanco, R. E. (1978). "Radiological impact of airborne effluents of coal and nuclear plants". *Science* 202 (4372): 1045-50.M http://web.ornl.gov/info/reports/1977/3445605115087.pdf

ONR (2011). *Japanese Earthquake and Tsunami: Implications for the UK nuclear industry, interim report*. 18 May. http://www.onr.org.uk/fukushima/interim-report.pdf

ONR (2014). *A Guide to Nuclear Regulation in the UK*. http://www.onr.org.uk/documents/a-guide-to-nuclear-regulation-in-the-uk.pdf

Poortinga, W., M.Aoyagi and N. Pidgeon (2013) "Public perceptions of climate change and energy futures before and after the Fukushima accident: A comparison between Britain and Japan." *Energy Policy* 62 November: 1204-1211.

Sovacool, B. K. (2008). "The costs of failure: a preliminary assessment of major energy accidents, 1907-2007". *Energy Policy* 36 (5), May: 1802-20.

Sovacool, B. K.(2010) "A critical evaluation of nuclear power and renewable electricity in Asia". *Journal of Contemporary Asia* 40 (3), August: 393-400.

UNSCEAR (2013). *Report of the United Nations Scientific Committee on the Effects of Atomic Radiation, Sixtieth session*. 27-31 May. http://www.unscear.org/docs/GAreports/A-68-46_e_V1385727.pdf

WHO (2013). *Health Risk Assessment from the Nuclear Accident After the 2011 Great East Japan Earthquake and Tsunami*. http://apps.who.int/iris/bitstream/10665/78218/1/9789241505130_eng.pdf

WHO (2014). *Seven Million Premature Deaths Annually Linked to Air Pollution*. http://www.who.int/mediacentre/news/releases/2014/air-pollution/en/

11

A change of ownership
(2012-13)

The rising cost of the EPR

The nuclear programme had survived the Fukushima disaster. DECC's carbon plan had reaffirmed the central role of nuclear in meeting the government's long-term carbon cuts, and the government's reforms of the electricity system, planning and regulation were all proceeding. But there was a new problem: the economics of the project were deteriorating fast. The cost of the Evolutionary Power Reactor or European Pressurized-Water Reactor (EPR) was rising just when EDF's ability to finance it was worsening.

The world's first EPR to start construction, Olkiluoto 3 in Finland, was falling further behind schedule. Construction started in 2005, and was to complete in 2009. By 2009, however, the project had a 2012 target and was $2.4bn over budget. By mid-2012 the estimated completion date had slipped to 2014. In July 2014 the Finnish electricity company TVO announced that the station wouldn't hit that target either and admitted it didn't know when it would be finished. TVO, the buyer of the plant, and the Areva Siemens consortium building it, each blamed the other for the delays.

Being the first EPR built, the uncertainties that inevitably accompany a new, complex project could, perhaps, be forgiven. The second EPR was in EDF's home market, at Flamanville in France. Flamanville-3, the third nuclear power reactor to be built on the site, started construction on 6 November 2007; it was expected that the 1,600 MW reactor would take 54 months to complete. The project was managed by EDF itself, with Areva again providing the main nuclear designs and expertise. The Italian electricity company Enel took a 12.5% stake in the project. Whatever problems were afflicting the first ERP in Finland, one might have hoped that lessons would have been learned for the second one.

In January 2010 *Le Figaro* reported the project was running two years behind schedule, which EDF denied (*Le Figaro*, 2010). But in July 2010 EDF confirmed that the station would not be ready until 2014 and would cost €5bn to build, about 50% more than originally expected. In August 2010 the French nuclear regulator found problems with welds in the steel containment liner and accused EDF of being slow in detecting "inferior weld quality" (Bloomberg, 2010).

In July 2011 EDF announced that the cost had risen to €6bn with completion now delayed to 2016. On 3 December 2012 EDF raised the estimated cost to €8.5bn. Enel pulled out, exercising its contractual right to compensation from EDF of €613m (£470m) plus accrued interest (Enel, 2012).

To be badly behind schedule on the first EPR might be considered unfortunate. To be failing on two EPRs looked rather as if there was something wrong with the design. There was some comfort from China, where Areva had started constructing two more EPRs at Taishan, Guangdong province, in November 2009 and April 2010. These were to be owned 70/30 by China Guangdong Nuclear (later China General Nuclear CGN) and EDF. These stations were widely reported to be on schedule and on budget, despite having a shorter construction schedule than in Finland or France. There would naturally be some benefits from being the third and fourth EPRs to be built, as lessons from the first two could be learned. And China's civil engineering expertise was built on a programme of infrastructure investment that was far larger than any other in the world. So perhaps the Chinese could build the EPR on time?

Alas, despite a steady flow of reassuring messages, it seemed even the Chinese could not build this complex machine as intended. There were reports of problems with the concrete pouring (as there had already been at Olkiluoto 3 and Flamanville 3). Areva proudly announced that the steel interior pressure vessel had been successfully fitted in June 2012 (Areva, 2012). But behind the scenes it was becoming clear that the construction schedule would not be met. In 2014, when the first reactor was due to be completed, it emerged that it was running at least two years late, which was something of an open secret in the nuclear world but was not officially reported.

In considering the economics of the twin EPRs at Hinkley Point C (HPC) in early 2012, EDF was more aware than anyone (except perhaps

Areva) that its new reactor was looking like a very troublesome beast. In most industrial processes one can count on a learning-by-doing effect: the first time is the hardest and most costly, but later versions of the same process should improve. This is a well-documented and very important economic principle. But there is scant evidence for it in the long history of nuclear power plant construction, even in France, which has built more reactors than any other country. The reasons are debatable but one is that nuclear power stations are largely built on site and each one is slightly different. There is always a tendency to add improvements, for technical or safety reasons, which frustrates the constructors' ability to get better at standardized processes. And the EPR is one of the most complex objects ever built.

EDF's deteriorating balance sheet and share price

Just as the HPC project costs looked likely to be far higher than originally expected, EDF's ability to handle this rising investment was looking increasingly doubtful. At the end of 2007, when stock markets generally were high, EDF's market capitalization was about €158bn (£115bn). Market capitalization represents the market's value of the equity of the company, the interests held by all the shareholders. It is calculated as the stock market price of each share multiplied by the total number of shares in issue, which in EDF's case is 1.86 billion (of which 84% are owned by the French government).

By the time that EDF completed its purchase of BE in January 2009, EDF's shares had dropped from their peak of €85 to only €35, although most of the decline was due to the worldwide fall in stock markets caused by the global financial crisis, which started to abate in early 2009. At that stage EDF's market value was €58bn.

A market value of around £50bn still represents a company with substantial financial resources, including the ability to mobilize funds for a major new investment programme. Nobody at that time doubted that, if it wanted to, EDF could, on its own, fund at least two new nuclear power stations in the UK. The government estimated in the 2008 white paper that a 1,600 MW reactor would cost about £2.8bn, so two reactors at HPC would cost £5.6bn spread over about eight years. A company worth £50bn could surely afford an extra annual investment of no more than £1bn? It might choose to bring in partners but that would not be to help with funding.

But, as Figure 11.1 shows, EDF's shares performed poorly in the next few years. This was partly because all European utilities shares did

Figure 11.1: EDF share price 2008-15 (€)

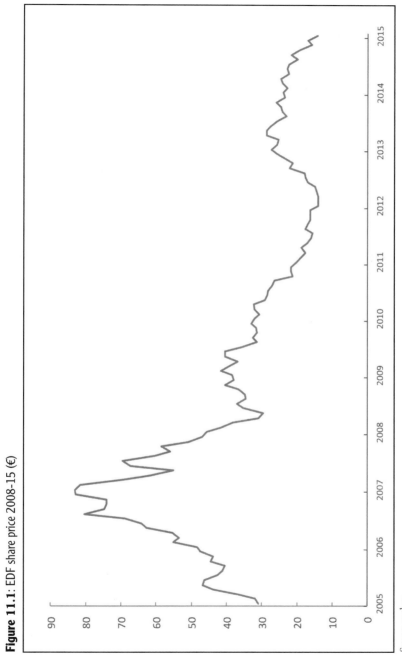

Source: author

badly (reflecting lower power prices). EDF, however, did worse than average, partly because of the Fukushima disaster (which threatened the prospects for EDF's international new nuclear build strategy), but mainly because of its home market, France, 2009 was a bad year for EDF's previously reliable French nuclear fleet. A series of strikes and then some breakdowns in the steam generators at several power stations led to a fall in nuclear output. In October 2009 EDF had to import power from abroad for the first time since the 1980s. The company's low-risk reputation was damaged.

And then there was the rising regulatory risk in France. EDF enjoyed a de facto monopoly in its home market but this was increasingly at odds with European Union rules on competition. The EU required all member states to open their electricity and gas markets to new entrants and make sure that the incumbent operators, which were mostly the former state-owned monopolies (like the CEGB had been in the UK) didn't unfairly block them. The goal was lower power prices and higher efficiency.

There were two problems with implementing this policy in France. First, it was very difficult to enter a market dominated by EDF, which controlled the vast majority of generation. In theory that generation capacity could be sold off to make two or more separate competing companies, as had happened in the UK. But splitting the nuclear fleet was a much more complicated problem than selling off individual coal or gas stations.

The second problem was that the French government was, to say the least, conflicted. It owned 84% of EDF and anything that damaged the company's financial standing also hurt the government. Moreover, EDF's leading role in the nuclear "Team France" gave it protection.[1] So for years the French government had dragged its feet on implementing the electricity competition policy. But even France complies with EU rules eventually. In 2010 the French parliament passed a new electricity law that would reorganize the electricity markets and provide a way for EDF to sell some of its power to competitors at a fair economic price. Quite what that price would be was unclear even in early 2012, but the uncertainty had taken its toll on the EDF share price. Investors were further worried when in April 2011 the tariff on the regulated part of EDF's income was raised by

---

[1] "Team France" was an informal name for EDF, Areva and the French engineering company Alsthom, which collaborated on French and international nuclear projects with support from the French government.

only 1.7% in what was seen as a deliberately political move ahead of the 2012 presidential election.

At the start of 2012 EDF's shares were down to €19 and its market value was only €35bn (£29bn). When presenting its 2011 annual financial results in February 2012, EDF confirmed that it was making progress with HPC and expected to make a financial decision in the second half of the year (EDF, 2012). It didn't give a figure for what the investment would now be but it must have had in mind a much larger figure than £5.6bn. When EDF signed its preliminary deal with the British government for a 35-year power price guarantee 18 months later, the cost had risen to £12bn, on top of some £2bn already invested in the project.

For a company worth £29bn to be contemplating an investment of nearly half its market value is a serious proposition. On top of that, EDF was facing a large increase in capital spending to refurbish its 58 domestic French nuclear power stations. These were on average 22 years old. A post-Fukushima report from the French nuclear regulator, the ASN (Autorité de Sûreté Nucléaire), in January 2012 confirmed that the stations were safe but that EDF must spend more to increase their protection against natural catastrophes, including extra back-up diesel generators, the creation of "bunkered" control rooms, better emergency management procedures and works for flood prevention. EDF had previously been telling stock market analysts it expected to spend some €40bn on refurbishment; these extra works would add a further €10bn.

The prospect of all of this extra investment made investors worry about EDF's financial strength, as measured by its level of debt. EDF, having enjoyed a protected quasi-monopoly position in France together with a large regulated business that is not much exposed to either power prices or competitive risk, had always been able to run with a high level of debt. But by 2012 investors were becoming concerned that the debt was growing just when the company's riskiness was increasing.

By 2012 investors were looking for reassurance that EDF could manage its debt and its position with the French government. They were not much concerned about the prospect of the company investing £14bn or so in the UK, partly because it seems many of them didn't believe this would ever happen. In meetings with stock market analysts, the company dutifully kept them abreast of its UK plans, but the analysts showed no interest. There were virtually no questions on the UK new nuclear strategy until 2013 when EDF announced the deal with the UK government.

EDF loses its partner

EDF did have a partner for HPC: the company Centrica, the largest retailer of electricity and gas in the UK. Throughout 2012 there were rumours of Centrica's growing doubts about the project. In April 2012 Centrica threatened to pull out of nuclear unless there was more certainty over future prices and clarity on energy policy. This seems to have been code for more assurance that the project costs could be contained and a government-guaranteed future power price would assure an acceptable return. A Centrica source told the *Financial Times*, "Nuclear power stations cost £6bn each and we must know what the return is going to be on that kind of investment. If we don't get the right answers, we won't proceed" (*Financial Times*, 2012a). This quotation suggested that EDF and Centrica were already thinking that HPC, which is in effect two stations, would cost £12bn in total. Centrica reaffirmed its official line that it hoped to reach a financial decision by the end of 2012.

Internally EDF had concluded that, if Centrica did pull out, it couldn't manage the whole HPC project on its own: even spread over eight years, the investment would be too large. EDF started to look for external partners who could take a share of the financing and of the risk. At the end of July 2012 EDF confirmed publicly that it was looking to cut its 80% stake in the Hinkley project by bringing in new partners. It also reported a further substantial increase in its debt, arising from a tax to pay for renewables in France (*Financial Times*, 2012c).

Meanwhile EDF was also pushing the UK government harder to improve the terms of the future power price guarantee. In line with the electricity market reform (EMR), which was still proceeding, EDF was negotiating a long-term "contract for difference" (CfD) – a fixed price guarantee for the power the station would produce. The Committee on Climate Change (CCC) had recommended this be set in an auction, but there was only one bidder, so the price was negotiated in repeated meetings between the government and EDF.

In 2012 the public was treated to a stream of newspaper stories about what that future price might be; sometimes it appeared that the government and EDF were negotiating through briefings to journalists. As EMR slowly wound its way through parliament, ever-increasing figures appeared in the press. On 23 July a person "close" to the negotiations told the *Financial Times* that a CfD of £100/MWh was being discussed. This was double the current wholesale power price and caused widespread alarm and a strong case of "I told you so" from those who had always argued that nuclear would not be economic.

The same day, the House of Commons Energy Select Committee, which was conducting pre-legislative scrutiny of the energy bill published on 22 May, designed to allow the government to make long-term price promises if it wished, concluded that such promises were currently unworkable because the Treasury wouldn't back them. It was particularly annoyed that the Treasury hadn't even sent a witness to its investigation. The committee had held a roundtable meeting with interested parties in June, which revealed widespread scepticism that the government's electricity reforms would unleash the £100bn of private investment that was needed both to meet the decarbonization and renewable goals, and to simply keep the lights on (House of Commons, 2012). The reforms were described as too complicated. Investors were concerned that DECC, the author of the reforms, did not have the full backing of the Treasury, raising the risk that future price promises wouldn't be watertight.

The nub of the problem was that the government was proposing long-term price guarantees for renewable and nuclear power. But the state doesn't actually buy wholesale power, so who was the contract to be with? The Treasury was adamant that it couldn't be the state itself. This would run the risk of breaching EU state aid rules and might add to public sector debt, which was already a serious problem. So DECC had come up with a "synthetic" counterparty, which impressed nobody and confused everybody. No one would invest on this basis.

Part of the problem was that the DECC and Energy Secretary Ed Davey (who had taken over in February from Chris Huhne, who had resigned abruptly for personal reasons) were giving assurances to consumers that future power prices would not be raised by the cost of decarbonization and renewables. The argument went as follows. Current power is mainly sourced from fossil fuels. Future fossil fuel prices are likely to be much higher, driving up the cost of the power produced from them. So the new renewable and nuclear energy supplies, although they look more expensive *now*, will be cheaper than fossil fuel power will be *in future*. So the policy not only will save the environment but will save the customer money. Davey argued further that energy efficiency improvements would cut the amount of power actually used per household, another reason for thinking that future power bills would not actually rise.

Investors saw this as coming dangerously close to a political commitment. Nobody knew whether future fossil fuel prices would go up. (As we now know, oil prices halved in the second half of 2014, against most forecasts.) If fossil fuel prices *fell* in future, then the cost of the government's energy strategy to consumers would be huge, compared with the status quo of sticking with coal and gas. In that case, the government would be under great pressure to break the contracts that it had signed in the past, leaving investors high and dry. This was not a theoretical debate. Spain, having initially provided very attractive prices for investment in solar power, had reneged on its contracts, wrecking the financial returns on many investments made in good faith on the promise of higher prices. The UK might do the same.

Through the summer, ever higher CfD prices for nuclear were rumoured. In August the head of EDF Energy (the UK subsidiary of EDF), Vincent de Rivaz, told the *Telegraph* that reports that EDF was seeking a price of £165/MWh were "utterly rubbish". He was happy to confirm that it was seeking a price of less than £140/MWh, in other words nearly treble the current price. He cited the previous year's CCC estimate of new nuclear costs of £100/MWh before then saying that things had moved on and that figure was now too low. He did at least reassure the public that EDF was not asking the government to take on the construction risk, given its confidence that it would deliver "on budget and on target". He also downplayed the consequences of a potential exit by Centrica from its 20% stake in the project, saying that other investors would be attracted (*Telegraph*, 2012).

A few days later it was reported that EDF was seeking Chinese partners for the Hinkley investment (*Financial Times*, 2012b). EDF refused to comment. The reference to partners, plural, suggested conversations with both CNNC (China National Nuclear Corporation) and CGN (China General Nuclear). The Chinese would turn out to be critical to the project (see Chapter 15). But EDF wasn't only looking for funding; it wanted to avoid a partner in China turning into a major competitor in the UK.

Meanwhile Hinkley Point continued to move towards the finishing line, even though it wasn't clear how it would be paid for. On 26 November the Office for Nuclear Regulation (ONR) granted a site licence to the project company, NNB Genco (NNBG), owned 80% by EDF and 20% by Centrica. This represented the culmination of more than three years and the equivalent of 6,000 days spent engaging with and assessing NNBG's "suitability, capability and competence to hold a nuclear site licence" (ONR, 2012). It was the first nuclear site licence to be granted in the UK for 25 years.

However, it didn't yet amount to permission to start construction. That required generic design approval (GDA) for the EPR, which came on 13 December, also from the ONR. That left only planning permission, which had to come from the Secretary of State, and final permission from the Environment Agency for radioactive emissions and other impacts. The GDA was very welcome news in France, since it meant that the EPR reactor had passed the stringent UK assessment, even if it were never actually to be built. EDF Energy's Vincent de Rivaz welcomed the news and said that the main EDF board would take a final decision in 2013 (missing the previous deadline for a decision of the end of 2012).

On 4 February 2013 Centrica finally withdrew from the project, writing off £231m. EDF now owned 100% of the project company. Later in the year Centrica's shareholders heard the reasons for the decision at their Annual General Meeting: costs had "rocketed hugely" their Chairman said. CEO Sam Laidlaw said that the project had stretched from an original 4-5-year schedule to 9-10 years, later clarifying that he meant all aspects of the project, not just construction. But he left a clear impression of a project in deep trouble, way behind schedule and far over budget.

Waste disposal problems

Meanwhile there was an unexpected setback for another aspect of the government's nuclear policy: the long-term storage of the waste. In 2008 the government had published its policy of storing long-term nuclear waste in a geological disposal facility (GDF) described as:

> A highly-engineered facility capable of isolating radioactive waste within multiple protective barriers, deep underground, to ensure that no harmful quantities of radioactivity ever reach the surface environment. The development of a GDF will be a major infrastructure project of national significance. It will provide a permanent solution for the UK's existing higher activity radioactive waste (including anticipated waste from a new build programme).

> (DECC, 2014)

Several other nuclear countries were intending to build GDFs, but the obvious difficulty was where? The British approach was to invite local communities to contact the government and see what would be involved, including financial payments. Allerdale Borough Council and Copeland Borough Council, both in west Cumbria, near to where most of the waste was stored at Sellafield, had expressed interest. On 30 January 2013 they took individual decisions on whether to participate in the next stage of the siting process. This was not a decision on whether to host the facility, but on whether to carry out further work to identify and assess potentially suitable sites in west Cumbria. Allerdale's and Copeland's council executives both voted in favour of further participation. But the DECC had previously agreed that the decision would need to be supported by Cumbria County Council, the next layer up in the local government structure. On the same day Cumbria voted against.

Cllr Eddie Martin, Leader of Cumbria County Council, explained why the 10 members of the County Council cabinet had voted against:

> Cabinet believes there is sufficient doubt around the suitability of West Cumbria's geology to put an end now to the uncertainty and worry this is causing for our communities. Cumbria is not the best place geologically in the UK – the Government's efforts need to be focused on disposing of the waste underground in the *safest* place, not the *easiest*.
>
> (Cumbria County Council, 2013)

There was no doubt that Cumbria had been seen as the easiest place because that was where Sellafield's "nuclear dustbin" was located and a degree of local support for nuclear-related activities there was expected. The borough council support suggested there was such support, but the county disagreed, citing a National Audit Office (NAO) report from 2012 that criticized the site for posing a "significant risk to people and the environment" because of the deteriorating conditions of radioactive waste storage facilities.

The only other local authority which had appeared interested was Shepway District Council in Kent, which had also taken soundings from local residents. But it subsequently decided against making a formal expression of interest.

So the government was close to approving a new nuclear power station but had nowhere to store the waste. This seems like a serious obstacle, but the government policy in the original nuclear white paper of January 2008 was carefully and somewhat flexibly worded:

Our policy is that before development consents for new nuclear power stations are granted, the Government will need to be satisfied that effective arrangements exist or will exist to manage and dispose of the waste they will produce.

(BERR (2008) 99)

So long as the government was satisfied that effective arrangements *would* exist, it could grant planning permission even if those arrangements were not actually in place. This turned out to be an important loophole, since otherwise there were likely grounds for a legal challenge to the entire new nuclear programme. The new stations would be built with interim storage facilities that would store their entire lifetime waste, for at least a hundred years, so in truth there was no practical urgency about the GDF. It would be remarkable, however, if the government, having paid so much attention to all the other aspects of nuclear policy, were unable to say just where the waste would ultimately be stored.

Hinkley Point C finally gets initial agreement

Hinkley received planning consent on 19 March 2013. Secretary of State Ed Davey told the House of Commons that he expected the remaining outstanding matters, including the radioactive waste decommissioning fund and the power price guarantee, to be settled "shortly". He finished his statement by describing his decision as "an important milestone in the process to decarbonise our electricity supply and economy" (HM Government, 2013).

A milestone it was but not quite as close to the final destination as expected. Just as the UK was approving the increasingly costly EPR, doubts surfaced from the unlikely source of EDF's Paris headquarters. The French press quoted EDF's Head of Production and Engineering, Herve Machenaud, as saying that EDF had lost its dominant position in design and construction and needed to develop a new generation of reactors smaller than the huge and apparently hard-to-build EPR (Reuters, 2013).

In March 2013 the negotiations were becoming increasingly tense. After a year during which CfD prices of £100/MWh and more had been quoted in the press, the Treasury was reported to have become involved, alarmed at the cost of nuclear electricity. It had insisted on a price of £80, equivalent to a return of 8% rather than the 10% figure that EDF had been seeking. EDF said it might be forced to start laying off people at the Hinkley site. EDF Energy CEO Vincent de Rivaz described the talks as "very challenging". Other electricity companies in the UK were warning the government not to commit to such an expensive power contract. But this all had the appearance of further negotiation by media coverage (*Telegraph*, 2013b).

EDF's group CEO Henri Proglio gave yet another deadline when he said in July that the company would take a decision on Hinkley by the end of the year. The Energy Minister, Michael Fallon, was reported to have been in China, visiting CGN, one of the rumoured investors. More interestingly he was reported to have indicated that "the government was hopeful that CGN might lead the construction of subsequent UK nuclear projects". Was this another negotiating ploy, putting pressure on EDF to sign or risk losing future nuclear investments to the Chinese? There was no reason in principle why CGN couldn't lead future projects rather than being a minority investor, as was rumoured for Hinkley. But that would be a radical new step for British energy policy (*Telegraph*, 2013a).

The British political consensus in favour of new nuclear power, already strong, grew firmer in September 2013. To everyone's surprise, including their own, Liberal Democrat party members voted at their annual conference to support the building of new nuclear power stations, after an impassioned speech by the Liberal Democrat Secretary of State for Energy and Climate Change, Ed Davey. This was a historic reversal of their long-standing opposition to nuclear power. The coalition agreement, noting that this was one subject that divided the party from the Conservatives, had allowed the Lib Dems to remain opposed or neutral on this issue, but that hadn't in practice slowed down policy. Now the coalition government was even more united.

Not everyone was pleased. Craig Bennett of Friends of the Earth (FoE) warned that "Ed Davey is deluded if he thinks new reactors can go ahead without public subsidy" (BBC News, 2013). It was soon to emerge that Davey wasn't in fact deluded, though the nature of the "subsidy" was going to be contentious.

On 21 October the British government announced that it had reached "initial agreement" on HPC – an agreement that was not yet contractually binding but was otherwise ready for implementation. Davey's statement listed the benefits of the investment agreement in the following order: i) up to 25,000 jobs during construction and 900 permanent jobs; £16bn investment in the UK economy; ii) a promise by EDF that £100m would go into the economy every year; and

iii) Hinkley would be a stable source of low-carbon power for 7% of the UK's electricity needs by 2025 (DECC, 2013).

The full details are discussed in Chapter 13. But Davey's statement confirmed the rumours of Chinese involvement. EDF had announced the "intent" of the two leading Chinese nuclear companies to invest in the Hinkley project as minority shareholders. EDF had apparently found its funding and the British government had apparently concluded a deal that would mean the UK's first nuclear power station in 25 years would start construction. But at what price?

Conclusion

The October announcement provided the basis for EDF to sign contracts for the construction of the now £16bn station. But the terms of the deal were to prove controversial, not just for the British electricity consumers who would ultimately pay for it, but for the European Commission (EC). The British government was intending to sign possibly the longest commercial energy contract in history for a price of power double the current level. How could that be squared with "no public subsidy"?

REFERENCES

Areva (2012). *Press Release: China. The construction of the EPR reactor unit 1 at Taishan takes a major step forward with the installation of the vessel.* 5 June. http://www.areva.com/EN/news-9361/china-the-construction-of-the-epr-reactor-unit-1-at-taishan-takes-a-major-step-forward-with-the-installation-of-the-vessel.html

BBC News (2013). "Lib Dem vote backs nuclear power plants". 15 September. http://www.bbc.co.uk/news/uk-politics-24100833

BERR (2008). *Meeting the Energy Challenge. A white paper on nuclear power.* January 2008. Cm7296. https://www.gov.uk/government/uploads/system/uploads/attachment_data/file/228944/7296.pdf

Bloomberg (2010). "EDF has welding problems at Flamanville reactor, watchdog says". 30 August. http://www.bloomberg.com/news/articles/2010-08-30/edf-has-welding-problems-at-flamanville-epr-reactor-french-watchdog-says

Cumbria County Council (2013). *Managing Radioactive Waste Safely (MRWS).* http://www.cumbria.gov.uk/planning-environment/nuclear/mrws.asp

DECC (2013). *Agreement reached on new nuclear power station at Hinkley.* 21

October. https://www.gov.uk/government/speeches/agreement-reached-on-new-nuclear-power-station-at-hinkley

DECC (2014). *Implementing Geological Disposal*. URN 14D/235. July. https://www.gov.uk/government/uploads/system/uploads/attachment_data/file/332890/GDF_White_Paper_FINAL.pdf

EDF (2012). *Annual Results 2011*. Appendices. 16 February. http://shareholders-and-investors.edf.com/fichiers/fckeditor/Commun/Finance/Publications/Annee/2012/2011EDFGroupResultats_annexes_Vfinal_va.pdf

Enel (2012). *Enel and EDF terminate their cooperation on EPR in France*. 4 December. http://www.enel.com/en-GB/media/press_releases/enel-and-edf-terminate-their-cooperation-on-epr-in-france/r/1656525

Financial Times (2012a). "Centrica threatens nuclear pull-out". 20 April.

Financial Times (2012b). "EDF looks for Chinese partners". 3 September.

Financial Times (2012c). "EDF open to partners on UK nuclear scheme". 31 July.

HM Government (2013). "Edward Davey statement on Hinkley Point C nuclear power station". 19 March. https://www.gov.uk/government/speeches/edward-davey-statement-on-hinkley-point-c-nuclear-power-station

House of Commons (2012). *Annex 1: Note from roundtable meeting*. Energy and Climate Change Select Committee. 23 July. http://www.publications.parliament.uk/pa/cm201213/cmselect/cmenergy/275/27515.htm#a54

Le Figaro (2010). "Le grand chantier nucléaire d'EDF prend du retard". 19 January. http://www.lefigaro.fr/societes/2010/01/19/04015-20100119ARTFIG00314-le-grand-chantier-nucleaire-d-edf-prend-du-retard-.php

ONR (2012). *ONR grants nuclear site licence for new UK power station*. 26 November. http://news.onr.org.uk/2012/11/onr-grants-nuclear-site-licence-for-new-uk-power-station

Reuters (2013). "EDF eyes development of new, smaller reactors – papers". 21 March. http://www.reuters.com/article/2013/03/21/edf-reactors-idUSL6N0CD1J820130321

Telegraph (2012). "EDF Energy puts price cap on Hinkley nuclear plant". 12 August. http://www.telegraph.co.uk/finance/newsbysector/energy/9471193/EDF-Energy-puts-price-cap-on-Hinkley-Point-nuclear-plant.html

Telegraph (2013a). "EDF eyes Hinkley Point nuclear decision 'by the end of the year' – again". 30 July. http://www.telegraph.co.uk/finance/newsbysector/energy/10211782/EDF-eyes-Hinkley-Point-nuclear-decision-by-the-end-of-the-year-again.html

Telegraph (2013b). "Nuclear future hangs in the balance as EDF talks reach 'critical' stage". 7 March. http://www.telegraph.co.uk/finance/newsbysector/energy/9914126/Nuclear-future-hangs-in-the-balance-as-EDF-talks-reach-critical-stage.html

PART IV

The world's most expensive
power station
(2012-14)

12

The Hinkley Point C project
(2013)

Hinkley Point C – the historical background

Hinkley Point is a headland on the coast of Somerset, on the Bristol Channel – the widening expanse of sea that emerges from the river Severn estuary and that is the historical boundary between England and Wales. It is an area of scientific interest and is popular with bird watchers and nature lovers. But its geography – access to the sea, close proximity to major transport links, connection to the national electricity transmission grid and location in the south of England – makes it an ideal location for nuclear power stations. Construction of the first station, a Magnox, started in 1957. An AGR, Hinkley Point B, started construction a decade later.

After construction started on the UK's first PWR at Sizewell B in 1987 the CEGB wanted to build an identical station at Hinkley Point, the first of five more PWRs across the country. In March 1988 Energy Secretary Cecil Parkinson announced a public inquiry. The inquiry took place against the backdrop of electricity privatization and the withdrawal of nuclear stations from the sale (Chapter 3). Despite local opposition the CEGB received planning permission in September 1990 for Hinkley Point C (HPC), but suspended any decision until 1994, pending the government's review of nuclear power. The CEGB's successor company, BE, decided in 1995 that building new nuclear would not help its privatization prospects and the permission lapsed.

The revived HPC project (2008)

Hinkley Point remained the most favourable site for a new nuclear power station. When new nuclear came back on to the policy agenda in 2007, it was at Hinkley that EDF first started to take seriously a new investment. This time there was a good prospect of local support. EDF started public consultations in October 2008, even before it formally took control of BE (which owned the site). In March 2009 EDF, having now bought BE, nominated the site, which was formally approved by parliament's passing of the National Policy Statement for Nuclear in July 2011.

EDF's proposals for the new nuclear facility encountered far less resistance than similar plans had done two decades before. Indeed, it is striking to compare the letters pages of the Bristol-based *Western Daily Press* in the 1980s with the mid-2000s. The earlier local opposition to new nuclear is mirrored by the more recent corresponding hostility to wind farms. The prospect of a new nuclear power station attracted only muted criticism, mainly in respect of the construction works and new transmission cables. There is very little evidence, however, of opposition to the new nuclear station itself. One possible reason for this is that the original Magnox station, Hinkley Point A, had closed in 2000 and the new jobs offered by the much larger HPC were attractive.

Construction and logistical facts

The "new" HPC project is more than double the size of the first version, involving two 1.6 GW EPR reactors, rather than a single 1.2 GW PWR. Hinkley would be one of the largest civil engineering projects in the world and by far the largest in the UK. The statistics are remarkable:

- The construction site will cover 175 hectares;
- The main earthworks require excavating 4 million cubic metres of earth, equivalent to 1,300 Olympic swimming pools;
- At least 3 million tonnes of concrete will be needed;
- The station requires 230,000 tonnes of steel reinforcement;
- 12,000 new trees will be planted;
- 25,000 jobs will be created over the whole construction period;
- In operation the twin reactors will eventually employ 900 people for 60 years;
- The stations will power 5 million homes;
- In operation it will avoid 10 million tonnes of CO_2 emissions per year compared with gas generation.

(EDF Energy, 2013a)

Figure 12.1 shows an artist's impression of the station.

Figure 12.1: Artist's impression of Hinkley Point C

Source: EDF Energy / © EDF Energy, 2011

The domes of the twin reactors are in the centre of the picture. The existing AGR, Hinkley Point B, is at the bottom.

The project will be at least comparable in scale to London's Crossrail project (currently the largest civil engineering project in Europe) and much larger than the London Olympics development. It will not, however, be among the largest power stations in the world, which are dominated by hydroelectric generation.

Hinkley does excel, however, in terms of its very high cost. The construction cost announced at the time of the government commercial deal in October 2013 was £16bn. The maximum full cost, including financing and contingencies, revealed by the European Commission (EC) state aid approval document (see Chapter 15), was £24bn. This would make HPC the most expensive power station, and possibly the most expensive single constructed object, on the planet.

The most expensive power station to date is China's Three Gorges project, which involved a huge amount of earthworks, a vast dam and lock for ships, plus the costs of relocating 1.3 million people. The total project cost came in at $37bn according to the official Xinhua news agency. But the infrastructure part of that was only $19bn.

The controversial project, completed in 2006, created 22.5 MW of generating capacity, about seven times that of Hinkley, which is estimated to cost considerably more.

Large infrastructure projects like HPC have an unfortunate tendency to get into trouble. The world expert on these "megaprojects" is a Danish professor at the University of Oxford, Bent Flyvbjerg. Flyvbjerg summed up the characteristics of such projects in a 2009 article:

- Inherently risky owing to long planning horizons and complex interfaces;
- Technology and design are often non-standard;
- Decision-making, planning, and management are typically multi-actor processes with conflicting interests;
- Often there is "lock in" or "capture" of a certain project concept at an early stage, leaving analysis of alternatives weak or absent;
- The project scope or ambition level will typically change significantly over time with a risk of higher costs;
- As a consequence, misinformation about costs, benefits, and risks is the norm throughout project development and decision-making, including in the business case;
- The result is cost overruns and/or benefit shortfalls during project implementation.

(Flyvbjerg, 2009)

HPC has most of these features, though it is fair to say that the scope has remained relatively constant. What makes Hinkley even riskier than usual is that it is a very "non-standard" technology: namely the fifth iteration of a new type of reactor, not one of which has yet been completed and all of which are, to varying degrees, over time and over budget.

HPC also involves a huge project management challenge: the central project team, run by EDF, has to coordinate several construction companies and equipment providers. EDF is also the project manager for the Flamanville C EPR under construction in France. An optimist might believe that the overruns and problems in France will have helped EDF to do a far better job at HPC. A pessimist might not.

The UK currently suffers from shortages both of skilled people and of other aspects of the "supply chain"; sourcing these adds to the cost of complex infrastructure projects. The engineering consultants Mott MacDonald, in a 2011 report for the CCC, estimated that this "congestion premium" might add 25% to the longer-term underlying construction costs of projects such as nuclear (but also offshore wind)

(Mott MacDonald, 2011). This cost premium might eventually reduce, though it could become worse if the UK started building two or more nuclear stations at once. Inevitably it is economically most damaging to the more capital-intensive forms of energy generation, such as nuclear and renewable energy, where the construction represents the largest portion of the total cost.

As one might expect, HPC is a matter of great discussion and enthusiasm in the British engineering world, representing the largest source of business in decades. The UK hasn't built a nuclear power station since Sizewell B in 1995. It hasn't built many other major civil engineering projects either, apart from the London Olympics, Heathrow Terminal 5, and Crossrail, due for completion in 2019. The conventional wisdom in the engineering world is that all of these have been managed well, with the Olympics coming in just under its revised 2007 budget of £9.3bn (the original 2002 budget of £4bn was unrealistically low – for example, it omitted any contingency allowance, underestimated the cost of policing, and overestimated the private sector contribution) (NAO, 2007).

Terminal 5 had the doubtful distinction of being preceded by the longest public planning inquiry in British history, at 545 days topping Sizewell B's record of 340 days. Apart from some teething troubles with the baggage handling equipment, it was seen as a success and came in on time and on budget, according to a House of Commons Transport Committee report (House of Commons, 2008). But Terminal 5 was a private sector project, in terms of both the customer (BAA, the owner of Heathrow airport) and the delivery (Laing O'Rourke, which is also a major contractor for HPC).

Flyvbjerg's work suggests that problems are more likely when megaprojects involve public sector customers, though there are cases of huge cost overruns in purely private sector projects too. A 2014 report from accountants Ernst & Young estimated that almost two-thirds of major oil and gas projects have run over schedule, with an average excess cost of 59% of budget (Ernst & Young, 2014). Crossrail, which does have a public sector customer, is still some way from completion in 2016, but an interim National Audit Office (NAO) report published in January 2014, about halfway through construction, was cautiously positive that the project was on time and budget (NAO, 2014).

With this recent encouraging track record, the UK engineering industry is confident about HPC. One reason is the increasingly advanced computer modelling available to the project managers, an area where the UK has established some leadership. "Business information modelling" (BIM) is the use of computer simulations in construction planning and management. Building on the use of computer modelling in design and manufacturing going back two decades or more, the current state of the art is 4D modelling. This means not only modelling the entire project in great detail but simulating the construction sequence, where time is the fourth dimension. This approach in theory allows an entire construction project to be rehearsed virtually so that there are no surprises when the actual construction takes place. This is particularly important with HPC, which involves a great deal of fabrication on site. It should allow each of the many sub-contractors to work off a single information set, with each stage seamlessly connecting to the next, avoiding the danger that one contractor is unable to do its work because another is in the way.

Yet experienced civil engineers are the first to point out that a megaproject like HPC is still essentially a massive human endeavour, requiring the coordination of many people, in different teams and companies. No modelling can entirely anticipate all possible risks and contingencies and when "stuff" happens it's down to the project management team to make things work.

Conclusion

One of the central points in Professor Flyvbjerg's work on megaprojects is the need for contracts that are clear, comprehensive and which allocate risk in an efficient way. In particular, those who bear responsibility for key decisions should be incentivized appropriately by bearing the costs of any error of failure. Which brings us to the financial aspect of HPC.

REFERENCES

Committee on Climate Change (2013). *Fourth Carbon Budget Review, Part 2*. The cost-effective path to the 2050 target. December. http://www.theccc.org.uk/wp-content/uploads/2013/12/1785a-CCC_AdviceRep_Singles_1.pdf

EDF Energy (2013a). *Hinkley Point C: An opportunity to power the future*. February. http://www.edfenergy.com/sites/default/files/edf-energy-hinkley-point-c.pdf

Ernst & Young (2014). *Oil and Gas Megaproject Overruns to Cost Industry More than US$500b*. January. http://www.ey.com/GL/en/Newsroom/News-releases/news-oil-and-gas-megaproject-overruns-to-cost-industry-more-than-us500billion

Flyvbjerg, B. (2009). "Survival of the unfittest: why the worst infrastructure gets built – and what we can do about it". *Oxford Review of Economic Policy* 25 (3): 344-67.

Flyvbjerg, B. (2014). "What you should know about megaprojects and why: an overview". *Project Management Journal*. April/May 2014. Available at SSRN: http://ssrn.com/abstract=2424835

House of Commons (2008). *The Opening of Heathrow Terminal 5*. Transport Select Committee. 22 October. http://www.publications.parliament.uk/pa/cm200708/cmselect/cmtran/543/543.pdf

Mott MacDonald (2011). *Costs of Low-Carbon Generation Technologies*. May. http://archive.theccc.org.uk/aws/Renewables%20Review/MML%20final%20report%20for%20CCC%209%20may%202011.pdf

NAO (2007). *The Budget for the London 2012 Olympic and Paralympic Games*. 18 July. http://www.nao.org.uk/wp-content/uploads/2007/07/0607612es.pdf

NAO (2014). *Crossrail*. January. http://www.nao.org.uk/report/crossrail-3

Nucleonics Week (1991). 21 March.

13

An outstanding deal for EDF?

The financial deal[1]

Figure 13.1 shows who bears each category of risk in the construction of Hinkley Point C (HPC). These risks are those generally involved in a nuclear power station investment.

Figure 13.1: Main types of risk in the Hinkley Point C project

Risk	Comment	Management
Construction risk	High, given EPR history	Investors take full risk
Power price risk	High in a deregulated market like the UK	Completely hedged for 35 years
Financial risk	Sceptical lenders ask high interest rates	British government provides guarantee
Operational risk	Low, assuming EPR works properly	Investors take full risk

Source: Author

EDF and the government were at pains to emphasize that the construction risk would remain with the investors. Given the disastrous record of the two European EPRs under construction, that seemed to leave a significant problem for the EDF-led consortium. But the high power price deal provided a form of insurance for EDF: if it were to experience overruns and higher costs, then it would be able to recoup some of these over the rest of the project's life. In effect, the power price deal compensated for the risk of a higher construction cost.

---

[1]   A more detailed analysis of the HPC deal is available in Taylor, 2016.

EDF seemed to have got rather a lot out of the British taxpayer. In case it appeared to have given away too much, the government had added a "gain share" clause which provided for an unspecified amount of benefits to the customer in the event that HPC was built *ahead* of schedule and *below* budget. But it left open the possibility that when the station started operating in 2023 or so, it would make a very high profit for its investors, prompting a political outcry.

The contract for difference (CfD)

The first component of the deal with EDF was the price promise: the government agreed to a 35-year contract for difference (CfD) for the power that HPC produced. A CfD is a common financial markets trading arrangement that provides the equivalent of a fixed price promise. Its substance is that if actual prices are higher than the agreed "strike" price then the recipient pays back to the contractor the difference. If the market price is below the strike price, the recipient gets the difference. So recipients receive, net, the agreed price for what they're selling, whatever path the actual market price takes.

The CfD would not be between the government and the HPC project company but with the National Grid. The CfD would be a private law contract, and so, having the normal protection of the courts, it could not be broken without compensation. National Grid would pass on the costs to the various electricity suppliers, who would in turn pass them on to customers. So the burden of the higher price would be shared across all electricity users.

When Secretary of State Ed Davey announced the price to the House of Commons, he first stated that it would be £89.50/MWh. But he then revealed that if EDF's second new nuclear power project at Sizewell C didn't subsequently go ahead then the price would rise to £92.50/MWh, reflecting the "first of a kind" costs at HPC. Given all the uncertainties, it was prudent to assume that Sizewell C wouldn't go ahead, so most commentators focused on £92.50 as the most likely price.

The wholesale market price of power had traded around £50/MWh during 2013, apart from a spike in March up to about £75/MWh. It was dropping towards the end of the year and in 2014 fell to nearer £40/MWh. So the HPC contract committed electricity customers to paying about double the market price of power from whenever the station started to operate. How could this be justified?

Ed Davey told the country that "Consumers will not have to pay over the odds for new nuclear. The price agreed for the electricity is competitive with the projected costs for other plants commissioning in the 2020s – not just with other low-carbon alternatives but also with unabated gas. As set out to Parliament in October 2010, and again in February [2013], new nuclear will receive no support, unless similar support is also made available more widely to other technologies."

Davey was assuming that other low-carbon electricity would be even more expensive than nuclear.

Indeed, given the 2008 Climate Change Act commitment to low-carbon electricity, HPC's price didn't look too bad. Figure 13.2 shows that at the time the deal was published, it was exactly in line with the average costs estimated for new nuclear by the Committee on Climate Change (CCC). In its update on the fourth carbon budget, published shortly after the HPC deal was announced, the CCC confirmed that the strike price was as expected, its figures being based on estimates produced by consultants that suggested new nuclear would come in at £85-100/MWh.

Figure 13.2: Hinkley Point C price compared with other generation sources

Generation type	**Price (£/MWh)**
Offshore wind (2013)	£150.00
Onshore wind (2013)	£100.00
Hinkley Point C CfD	£92.50
CCC average estimate of nuclear cost	£92.50
Hinkley Point C if Sizewell C proceeds	£89.50
Average wholesale market price (2013)	£50.00

Source: CCC (2013) Chapter 3; Energy Solutions, 2015

Figure 13.2 also shows that the other two main practical sources of low-carbon power were even more expensive at that point. Government price deals for onshore wind were in the region of £100/MWh and for offshore wind £150/MWh. So the government could argue that for low-carbon power Hinkley was far more cost effective than offshore wind, and would reliably last for 60 years, much longer than current experience with offshore wind turbines in a hostile North Sea.

However, the cost of renewables was set to fall faster than the CCC projections. The CfDs issued in 2013 were allocated by negotiation.

This process was criticized by, among others, the National Audit Office in 2014, which urged a more market-based approach instead. A new set of CfDs for renewable energy projects, announced in February 2015, was based on an auction, with project developers submitting bids for a price they thought would give them an acceptable return. The results were encouraging: as Figure 13.3 shows, offshore and onshore wind prices were significantly lower than the figures paid in 2013. Solar PV (photovoltaic) projects were winning contracts at much lower prices than previously expected, and offshore wind was attracting bids at £120/MWh, much lower than the £150 paid in 2013.

Figure 13.3: Results of competitive auction of CfDs for renewables, February 2015

Generation type	CfD (£/ MWh)	Capacity (MW)	
		Average project	Total of winning bids
Offshore wind	£120	581	1,162
Onshore wind	£83	50	749
Solar PV	£79	14	72

Source: DECC (2015)

Note, however, that the amounts of capacity were relatively small. Two offshore wind farms totalled 1,162 MW, about a third of HPC. The average onshore wind farm was only 50 MW. A major reason for proceeding with Hinkley was that it would deliver a very large amount of capacity in one go – 3,200 MW. Moreover, the relative "capacity" figures understate the benefit of HPC. Wind farms might typically produce at about 30% of their maximum theoretical capacity (since it is not always windy and they have to be closed for maintenance). Nuclear should produce at around 90% capacity, the shortfall from 100% being for maintenance and inspection shutdowns and a small amount of unplanned outages (assuming the plant works properly). To reach the equivalent of HPC's annual energy output would require a wind farm of nearly 10,000 MW capacity (ie three times the size of HPC).

The debt guarantee

The second part of the deal with EDF was a government guarantee on the loans raised by the project company, NNB Genco (NNBG).

Figure 13.4: Targeted Hinkley Point C project structure at October 2013

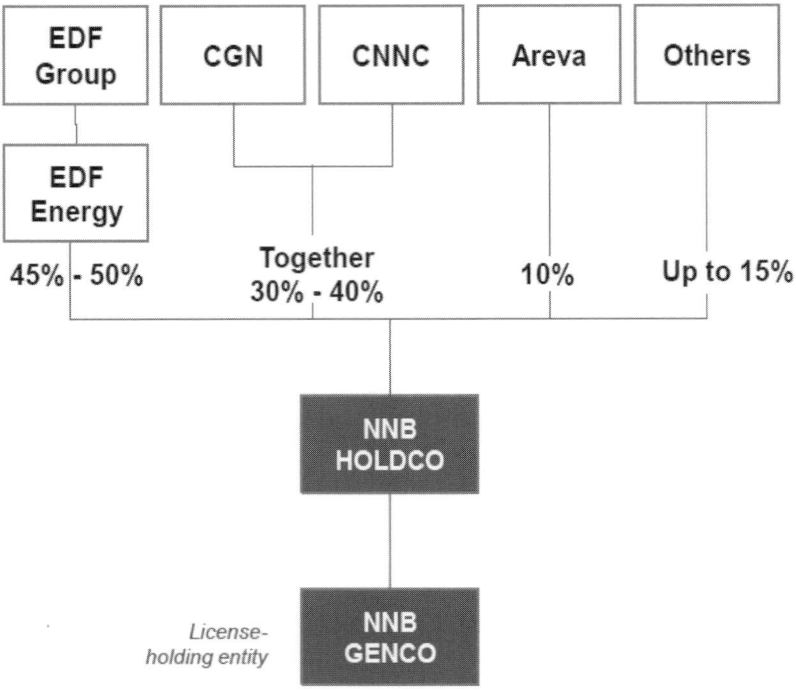

Source: EDF Energy, 2013

This company was a subsidiary of the UK company, EDF Energy, itself a subsidiary of the EDF group based in Paris. NNBG owned the HPC project and would sign the contracts for construction and operation, fuel procurement, insurance and financing. Why wasn't EDF doing this itself, as it had for all of its other power stations, nuclear and non-nuclear? There were two reasons.

The first was to provide the flexibility to bring in additional shareholders. EDF might have contemplated this before, but now it was essential because EDF couldn't finance the project on its own. The ownership structure of NNBG was completely flexible and EDF had announced that it expected eventually only to own a stake of 50% or less (Figure 13.4).

Since operating a nuclear power station requires a licence from the ONR, there must be a designated licence-holding company, which is NNB Genco. That in turn is owned by a holding company, NNB

Holdco, which is where the external shareholders come in. EDF's plan in late 2013 was eventually to sell down about half its equity ownership, with the balance being taken up by China General Nuclear (CGN), China National Nuclear Corporation (CNNC), Areva (the company that designed the EPR) and unspecified others. None of these had signed contracts, so the plan was just that, a plan. In particular, the participation of the Chinese investors was some way from completion. Indeed EDF's presentation to investors the day the deal was announced referred only to an "expression of interest" from Chinese investors (EDF, 2013). In addition, Areva's worsening financial position, mainly caused by its liabilities from the massive cost overruns at the Finnish EPR construction site, was to put its ability to invest in doubt.

The second reason for this structure was to insulate EDF from some of the project risk. It is quite normal in large infrastructure and energy projects for a group of companies to come together to bear the risk jointly, usually by creating a project company in which they are equity investors. That means they have limited liability. If anything goes wrong with the project, leading to a financial claim from other people (including a government), then those claims are on the project company, not the shareholders in that company. A funding structure like this is known as "non-recourse funding", because a claimant on the project has no recourse to the shareholders (no right to expect financial compensation from them).

If HPC somehow got into financial difficulty, its creditors could sue NNBG but not EDF or any other shareholders. The worst that could happen to the shareholders would be losing all of their investment in the project. But they would never legally be required to put in any further investment, though they might choose to in some circumstances. For example, if the project were to go badly over budget and exhaust the funds of the NNBG, the shareholders would not be obliged to put any more investment in. They might prefer to put some more funding in to complete the project, since a half-completed nuclear power station is worth very little; but they would at least have the option to write off their earlier investment and walk away.

While the holding company structure limited the financial risk to shareholders, this very protection would raise a doubt in the minds of potential lenders. There are two forms of financing: debt and equity. From a lender's point of view, HPC looks like a somewhat risky proposition, at least before the government's investment deal. There is the very real construction risk – the project might need many billions of extra pounds of cash to complete There is also the risk that the future selling price of electricity might fall short of the level needed to keep the project profitable. The CfD took care of the price risk but the construction risk, we were repeatedly assured, would remain with EDF, or more accurately with NNBG.

Would a sensible bank lend to NNBG, knowing the construction history of the EPR? In the cautious climate following the global financial crisis of 2008-09 banks were looking for safe investments like government bonds and secured loans such as mortgages, where the bank can take possession of the house if the borrower fails to pay.

Taking possession of a half-completed nuclear power station is not quite so reassuring. EDF, 84% owned by the French state, seemed like a good credit risk, because presumably it would be bailed out by the government if it got into trouble (it is, in effect, "too big to fail" like many banks). But the HPC project structure prevented banks from claiming any financial shortfall from EDF. Lending to the subsidiary NNBG made the loan riskier.

The upshot was that it was most unlikely that NNBG would be able to raise the billions of pounds of loans needed. That meant either the shareholders would have to put in a lot more equity, thus raising the overall financing cost of the project and lowering their expected returns, or they would need a credible guarantee for the banks. That credible guarantee eventually came from Her Majesty's Treasury; that is, the British taxpayer.

To be fair the Treasury had already set up a wider infrastructure debt guarantee scheme, for which HPC was potentially eligible; the problem that Hinkley faced was different from other infrastructure projects only in scale, not in kind. Total lending for UK infrastructure investment fell from £6bn before the financial crisis to £3bn in 2010. Given the urgent need for new infrastructure across many sectors, not just energy, the government decided it needed to step in to help, passing the Infrastructure (Financial Assistance) Act in October 2012. This act included the UK Guarantees Scheme ("the Scheme") covering energy, transport, health, education, courts, prisons and housing.

To avoid breaching EU rules banning state aid (see Chapter 15) the Scheme had to be arranged as a fee-based service, where the fee was supposedly set at a market rate. In theory, therefore, the project benefiting from the guarantee would pay a fair market price for it. (There is some doubtful logic in this: if the normal lending market were functioning properly, there would be no need for the Scheme in the first place.)

The Scheme works as follows. A project company pays a fee to the Treasury, which in turn promises to a bank or other lender to the project that the debt principal and interest will be paid, whatever happens to the project. The lender is, in effect, lending to the British government, which has not defaulted on a debt since the 17th century and is therefore regarded as extremely creditworthy. So the lender will charge the project an interest rate pretty much the same as if it were lending directly to the government itself, which is much less than the cost of lending to a risky project company. The project therefore benefits from a lower cost of borrowing, offset in part by the fee to the Treasury. Somebody trained in free market economics might ask how a project could be better off, since the fee to the Treasury ought to offset exactly the lower cost of funding. The project risk is unchanged; it's simply being redistributed from the lender to the Treasury. If the Treasury is pricing this risk correctly, then it ought to charge the project company an amount that fully reflects the risk that the lender is offloading. But then there is no benefit to the project company; it's simply paying a lower interest rate to the lender and a higher fee to the Treasury. The fact that this is *not* what happens shows that there must be some benefit to the company – a subsidy from the Treasury.

The NAO appears to have some sympathy for this argument, stating in a 2015 review of the Scheme that "There are no directly comparable market benchmarks for the Treasury guarantee fee because the guarantee is superior to commercial alternatives" and leading it to conclude that "We do not have full confidence in the reliability or completeness of market benchmarks used to measure actual risks to taxpayers" (NAO, 2015: "Summary").

The NAO's concern is with value for money for taxpayers: is the Treasury under pricing its guarantees to the benefit of the project investors? But the whole point of the Scheme is to provide some benefit to investors in projects of national importance, so even if the NAO is right to be anxious, that doesn't mean the Scheme is a bad idea.

The Scheme was launched with a closing date of 2014, later extended to 2016, reflecting the view that the lending market would eventually return to normal, rendering the Scheme no longer necessary. It was given a maximum limit of £40bn (excluding interest) and issued its first guarantee in January 2013. It later emerged that the HPC guaranteed amount could be as much as £17bn, by far the largest single guarantee. EDF expected that the guaranteed debt would meet 65% of the total project costs.

Indemnity against future government policy changes

EDF highlighted a third part of the deal to its investors, and did so rather more clearly than the government. This was a set of promises for compensation for "discriminatory changes in law" and for "political shutdown of the station by the authorities" unless it were safety related. EDF quite understandably wanted to be guaranteed against the risk of a German-style reversal of national policy on nuclear power or any other government action that might disadvantage nuclear relative to other electricity generators. This would also be something that the European Commission (EC) would later examine sceptically.

EDF's likely rate of return

The package of the CfD, debt guarantee and what EDF called "robust investor protection" would provide an estimated internal rate of return (IRR) of about 10% on the project, according to EDF's Group Senior Executive Vice President for Finance, Thomas Piquemal. What did that mean?

If I lend £100 and receive £10 interest plus my original £100 back, then I've made a 10% return. But if the £10 doesn't come back to me for two years, how do I value it compared with another investment that gives me £10 back in six months? The concept of the *internal rate of return (IRR)* is intended to capture the critical element of timing in the investment flows and returns of a project. It is a standard part of the corporate finance toolkit used by companies and taught in business schools (though it has its critics).

The IRR makes use of the concept of *discounting*. We discount the future when we value it at less than the present. All investment projects involve some financial investment now or in the short term, with an expected or hoped for return of cash flows in future. Since these inflows and outflows occur at different times, we need a way of comparing them, adjusted for timing. The IRR is one way of doing this and it also has the advantage of giving a percentage figure that can be compared with the return required by the investor. A simple decision rule could be: invest if the IRR exceeds the minimum required return.[2]

---

[2] Anyone trained in a good business school will know that the best way of

Figure 13.5: Indicative cash flows for Hinkley Point C (£ million)

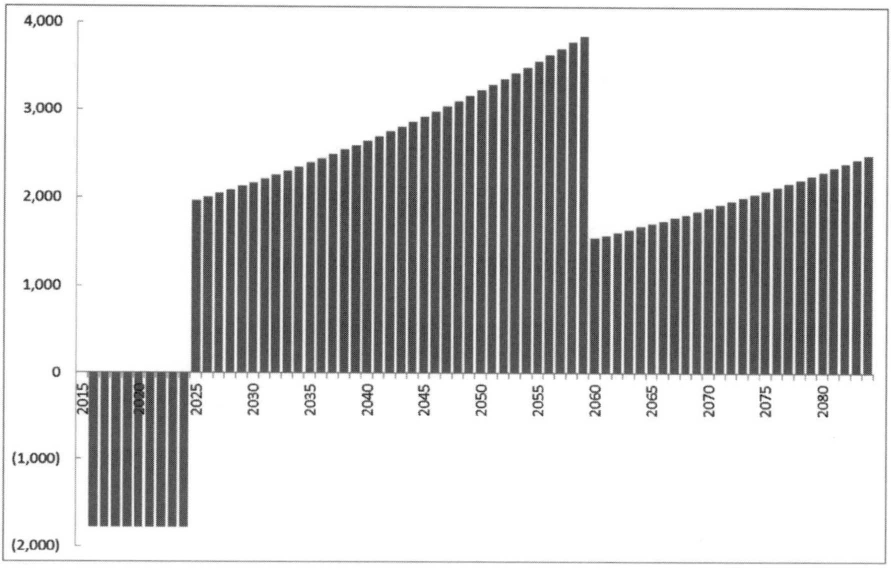

Source: Author's estimates

Nuclear power stations are among the longest-horizon investments made. HPC will take at least eight years to build, is designed to operate for sixty years and will then face defuelling and decommissioning that will stretch several decades further into the future. (The radioactive waste from the station will remain dangerous for centuries but that's the government's problem, because it is responsible for long-term storage.)

Figure 13.5 shows the pattern of cash flows for HPC: a period of investment during construction, a 35-year period of operation with the CfD guaranteeing inflation-based price increases, and a further 25 years at market prices.

There are 3 phases in the diagram. The first is the construction phase, which this exercise assumes will take 9 years. The chart shows an even spread of the cost (estimated at £16bn), which is unrealistic, but annual variations make very little difference to the overall economics. We then project profitable operation for 35 years, assuming 2% annual

---

conducting investment appraisal is to use another, similar approach called net present value analysis; companies, however, like IRR because it expresses the return as a percentage.

inflation, which is why the figures rise steadily. After 35 years the CfD expires and HPC receives whatever the market price of power is. We assume here, rather arbitrarily, that the market price is half that of the final year of the CfD, and that average prices continue to track inflation of 2% until the station closes after 2084 (see Taylor (2016) for more details).

(There is technically a fourth phase of the project, which is the cost of defueling and decommissioning the station after closure, a process taking up to 50 years. Those costs are to be met out of a fund that is built up during the station's operating life, in effect a pension fund for a retired nuclear power station. If the fund is correctly maintained then there should be no residual cash costs left after HPC closes in 2084. About £2 of the £92.50 strike price is to be allocated to the fund. If that sounds a small amount, it is because of the costs lie so far in the future; when discounted to the present they are far smaller.)

The operating cost, $23/MWh, is taken from the latest estimate of US nuclear power stations, which are of a different design and are older (Nuclear Energy Institute, 2015). HPC should have some operating advantages over the older stations but will face different costs for decommissioning and long-term fuel storage than in the USA. All of these are very hard to estimate from the outside so we've left the US figure (adjusted at £1.0 = $1.5) as the least bad guess.[3] We assume that HPC is available to produce power on average 91% of the time, which is in line with best practice among modern nuclear stations and is the factor assumed in the European Commission state aid assessment (European Commission, 2014). We assume UK corporation tax remains at 20%, which is clearly subject to change over such a long period. And we assume a fall in the average selling price of 50% in year 36, after the CfD expires.

Assuming these estimates are roughly accurate, what can we say about the value of this project to the investors? The IRR for the period until the end of the CfD turns out to be 9.6%, which is very much in line with the IRR estimated by EDF. That could be a complete fluke but it's likely that our calculations are broadly similar to EDF's. Essentially, investors have to ask themselves the question: if we commit to something like £16bn of construction costs over about nine

---

[3]   The NEI (2015) reports 2013 (the latest data available) US nuclear operating & maintenance costs of 1.51c/kWh plus fuel costs of 0.79c/kWh for a total operating cost of 2.3c/kWh, which is £15.3/MWh converted at £1 = $1.50

years and the plant runs efficiently over the next 35, is a 10% rate of return reasonable?

Is 10% the right return?

So is 10% a reasonable return for HPC? Corporate finance analysts estimate the required rate of return on a project using a method called the capital asset pricing model (CAPM). In finance jargon, the required return is known as the *cost of capital*: it is the "cost" in the sense that it is the percentage annual return needed to attract capital (ie funding) from investors, who have alternative uses for that capital and consider each project opportunity on the basis of its return versus its risk. The CAPM is known to be flawed and subject to a significant margin of error but it is widely used because it's convenient. It quantifies the risk using three components:

1. The percentage return on a risk-free investment – such as a bond from a creditworthy government, such as the US or UK.
2. Companies can go bust (unlike the UK government) so to compensate for the risk of investing in the stock market as a whole the investor needs an extra percentage, the *equity risk premium*.
3. A multiplier, called the *beta*, takes account of whether this investment is more risky or less risky than the overall market.

Total cost of capital is $(1 + (2 \times 3))$, expressed as a percentage.

Now we'll use CAPM to compare an average equity investment in the UK with HPC in several scenarios. This will not be exact, partly because the full information needed for a thorough analysis isn't available, and partly because estimating the cost of capital, although an absolutely central part of applied finance, is far from precise.

Average investment

We can define this as an investment having the level of risk an investor would get by buying a share in the overall London Stock Exchange (which is easily done through an index tracker fund). The CAPM analysis for this average investment is as follows:

1. The risk-free rate on UK government bonds was 2.7% in October 2013.
2. The equity risk premium represents an expectation of the future, so people will disagree the value, but here we assume 5%.

3. The beta adjusts for the riskiness, compared to average. Here (by definition) the risk is exactly average, so the beta multiplier is 1.

This gives us a required return/cost of capital of (2.7% + (5% x 1)) = 7.7% for a generic UK equity investment in October 2013. This is the benchmark to which we'll compare HPC.

HPC without the CfD

This would be very risky, because of both the construction risk and the risk of the future power price. The only reasonably firm assumption would be that the project would generate a lot of power for a long time at a fairly predictable cost. However, the revenue would be uncertain, reflecting the ups and downs of the UK electricity market, which in turn would be linked to other energy prices and to whatever the carbon price was then.

1. The risk-free rate is 2.7%, as above.
2. The equity risk premium is 5%, as above.
3. However, because this project is, say, more than twice as risky as the average project, we use a beta of 2.5.

2.7% + (5% x 2.5) = 15.2%, giving us a required return of 15.2%.

Even 15.2% might not be high enough for investors worried about a repeat of the overruns on the EPRs at Olkiluoto and Flamanville. They might require the sort of return expected in upstream oil and gas exploration, which can be 20% or more. Maybe the project would simply not be financeable at all. Conventional finance assumes that all risk can be "priced", meaning that there is a return high enough to compensate for any risk, but that might not be true for such a large and unusual project as HPC. But we can confidently say that without the CfD the project would require a far higher return than average.

HPC with CfD

What difference does the government-backed CfD bring? The revenue risk has been taken away, so long as the station generates electricity as expected. Few commercial investments offer a guaranteed revenue stream in real terms for 35 years into the future. But the construction risk still looms large. So let's assume that the project remains riskier than the average generic project but now with only a factor of 1.5 times, giving an overall cost of capital of **2.7% + (5% x 1.5) = 10.2%**. This is in line with the project IRR of 10% from our calculations and from EDF's presentation to its investors.

Note that the main remaining risk in the project is construction risk. EDF repeatedly assured the government and investors that it would bear that risk (or more accurately the NNBG holding company would). So the 1.5 risk enhancement factor that gets us to a 10.2% required return is really only that high because of the (non-trivial) risk that the construction goes over time and budget.

Replacing equity with loans increases the return

The 10% we have been discussing is the return on the whole project, ignoring how it is funded. What shareholders ultimately care about is the return on the equity they invest. The more they can fund with debt, the higher the return on the equity will be, though the risk is also higher.[4] Once we allow for the project being partly funded by debt then we can assume that the equity return will be higher.

So far we have assumed the project is entirely funded with equity, but that would be an unusually costly way of funding a project because equity is more expensive than debt. Banks and bond investors would be wary of lending during the construction phase, which is where the government debt guarantee comes in. It is not clear exactly how much of the project would be funded with debt, but we learned in 2014 that the total amount of guaranteed debt could total £17bn (see Chapter 15). That implies a high proportion of debt to equity in the project. How might that change the returns?

1. Assume the project replaces 65% of the equity with debt. We must increase the cost of equity because the deal is now highly levered and therefore riskier: if there is any problem with the project then the value of the equity is more likely to be completely lost. So we have raised the risk adjustment factor (the beta) to 2, which is pretty high. The cost of the equity capital is now **(2.7% + (5% x**

---

[4] Consider a simple example: a house bought for £250,000, which is now worth £300,000. If you bought it with a deposit of £100,000 plus a £150,000 mortgage, the return on your equity is (£300,000 – £250,000 = £50,000)/£100,000 = 50%. If you bought outright, your percentage return is only £50,000/£250,000 = 20%. By using a mortgage you get the same absolute return, £50,000, on a much smaller amount of your capital.

2)) = 12.7%, somewhat higher than the 10.2% without debt.
2. Next, consider the effect of the debt. Assume the interest rate payable is 5% instead of the 2.7% paid on government debt. (The project has a government guarantee so this seems a good deal for a lender, as they are sure of getting their money back.) Since debt interest is tax deductible, the cost after tax is 4% (ie 5% less 20% corporation tax relief on the interest).

The total cost of capital to the project is therefore 4% on 65% (ie the debt portion) of the money invested, plus 12.7% on 35% (ie the remaining equity portion), giving an average cost of 7.0% overall.

That is the effect of using debt, made possible by the government guarantee.

The surge in value when construction is complete

These estimates of the HPC project financing cost are taken from the point of view of investors at the start of the project, surveying its whole life, including the 9-year construction period.

However, once HPC is actually built and operating it becomes a relatively low-risk investment. Many investors might like a share in a project with a 35-year guaranteed revenue stream, with very little commercial risk. The risk factor on such a project would be less than the average, possibly quite a lot less. If we apply a risk factor (beta) of 0.8, in line with lower risk companies in the London stock market, the required return drops to only (2.7% + (5% x 0.8)) = 6.7%. (That's without leverage. If we replace some equity with debt, the required return is even smaller.) This gets us to the heart of a problem with the government's deal. It is a very inefficient way of covering the main risk of the project, which is the construction risk. Instead of explicitly covering or at least mitigating that risk, it provides a 35-year price promise that potentially overcompensates the investors for a risk that will have dropped out by the time the station operates. How else could the deal have been struck?

To take the construction risk completely away from investors the government could have: i) taken over ownership and sold the power station on completion; or ii) provided a full guarantee to the investors against overruns. Neither of these was politically likely, given how difficult it was for the government to wriggle round its earlier promises not to subsidize nuclear power. Providing a guarantee would almost certainly have been judged inconsistent with European state aid rules (see Chapter 15). Buying the project outright and then selling it on would have been difficult too. The only potential buyers would be those with suitable nuclear credentials and capable of getting a licence from the regulator. That might have ended up with only EDF as a credible buyer. So the government would have faced the challenge of selling a new nuclear power station with few potential buyers, and perhaps only one.

So there is a risk that the NNBG consortium manages to build the station on time and on budget and makes an excessive return for what would then be a relatively low-risk project for the next 35 years. This would be wonderful for the shareholders but politically embarrassing. Imagine the situation in 2026 after the first full year of operation of HPC. The holding company, not being public, would need to publish only the most basic accounts. But EDF, which is a publicly listed company, would need to show how much profit it was getting on its equity investment. That figure might well be around £1bn, assuming EDF has a half-share in the project.

If in 2026, on top of its probable underlying £800m (from its other UK operations), EDF reports an extra £1bn of profit, entirely attributable to the contribution of HPC, it will be rather hard to hide. Even more politically risky would be if EDF at that point were able to refinance the project by replacing the equity with cheaper debt (which would be an entirely rational and reasonable thing to do). The construction phase absolutely requires a lot of equity to take the risk that things go wrong. But once that construction risk has gone it would be easy to persuade banks or bond investors to lend to the project company, because the revenues are covered by a British government price contract. This new debt wouldn't be covered by the original debt guarantee, but it wouldn't need to be. The project would by now be a relatively safe one. Many investors would jump at the chance to make a 35-year investment at low risk because such opportunities are very rare and mainly confined to government bonds, which pay a very low return (at present).

What would this mean for EDF and the other investors? It would mean they could take a very large dividend in the form of withdrawing the no longer needed risk equity. This is not just a theoretical possibility: it is exactly what has happened in other past UK contracts where private investors refinanced the project after the construction risk dropped out.

Learning from the PFI refinancing scandal

The 1992 Conservative government under Prime Minister John Major introduced the Private Finance Initiative (PFI), a mechanism for financing new public sector infrastructure, such as schools and hospitals, with private funding. The PFI is a UK version of the wider concept of private public partnership (PPP) in which the customer is a public sector body but the delivery, financing and operation are met by private sector suppliers. The goal of the PFI under the Conservatives and then under the Labour governments of Tony Blair and Gordon Brown was ostensibly to improve the efficiency of both construction and operation, but a less publicly stated purpose was to keep the infrastructure funding costs off the government's balance sheet (though later changes in accounting rules stopped that).

The projects were typically in the tens of millions of pounds; sometimes in the hundreds of millions. Like HPC, they involved consortia of private companies that owned a project company. The project company funded the construction, which was the most risky phase of the project. After construction the project was much less risky and enjoyed a long-term (20 to 30 years) period of guaranteed income from the public sector customer.

The early PFIs became notorious for their high returns. In defence of the companies, these were somewhat pioneering and untested investments. But the returns were then increased, sometimes dramatically, by refinancing the projects after construction. Exactly as we have conjectured for HPC, these projects started with a high level of relatively costly equity during the construction phase but swapped it for cheaper debt when the construction risk had dropped out.

Figure 13.6: Highest rates of return on PFI project refinancing, after 30% sharing with customer (%)

Project	IRR before refinancing	IRR after refinancing
Debden Park School	15%	71%
Norfolk & Norwich Hospital	16%	60%
Bromley Hospital	27%	70%
Darent Valley Hospital	23%	56%

Source: HoC Committee on Public Accounts (2007) Fig. 1.

Early PFI projects were signed under a voluntary code of practice that required 30% of the refinancing gain, which is very likely to occur in any normal project, to be shared with the customer. Projects signed after July 2002 make this 50% (House of Commons, 2007). Yet the investor returns were still very high. The refinancing led in the most extreme cases to returns of more than 50% (see Figure 13.6).

The Norfolk & Norwich hospital deal in 2006 was particularly controversial, as the project was followed by a threat of 450 job losses, which were blamed in part on the cost of paying for the PFI. The project sponsors originally borrowed £200m and two years after completion raised this to more than £300m, allowing a £115m dividend to be paid. This was an early PFI hospital project with some hard-to-estimate risks. But that didn't stop MPs from criticizing it. The Conservative MP, Edward Leigh, Chairman of the Public Accounts Committee, described it as "the unacceptable face of capitalism" (*Guardian*, 2006). (However, some PFI projects lost money for their investors, so the average return achieved has been lower than the very high figures in Figure 13.6.)

On the plus side, 69% of PFI construction projects in 2003-08 were built on time and 65% were on budget. Of those delivered late, 42% were delivered within six months of the agreed time, and less than half experienced price increases (NAO, 2009).

The danger is clear, though. The government could find itself in a position where the investors in HPC, if it is completed on time and on budget, pay themselves a multi-billion pound special dividend, crystallizing a dramatically higher return on the project than was envisaged at the start.

There may be readers who are wondering where Her Majesty's Treasury was in all of this. The Treasury is the ultimate protector of the public purse. And even though in HPC the cost of the guaranteed power price will fall on the electricity customer, not the taxpayer, these are largely the same people.

In fact there were some wise heads who saw the dangers of the deal and insisted on a mechanism for sharing some of the benefits of the project with customers, if it went well. The government therefore included a "gain share" mechanism that provided for unspecified payments from the project company to customers (in the form of a lower CfD price) if the project turned out to be more profitable than expected, particularly if construction finished ahead of time and budget. The exact terms of that deal were not made public until the

EC revealed them and demanded that they be strengthened in favour of the customer (Chapter 15).

A City of London view

In case this analysis seems implausible, here is a view from a City analyst, whose job is to advise professional investors in pension funds and hedge funds on investing in the European utility sector. In a now famous report published on 30 October 2013, shortly after the HPC deal was published, Peter Atherton and Mulu Sun of Liberum Capital, an independent investment bank based in London and New York, described their reaction to the deal as "flabbergasted", which was the title of their report on the deal (Liberum Capital, 2013).

Peter Atherton is not just any utility analyst, having for several years been the most highly rated electricity analyst in the annual Extel and Institutional Investor surveys. He is one of the longest-serving analysts in London covering the electricity sector. His report concluded that "this looks likely to be an outstanding deal for EDF and its partners". He estimated that the investors could achieve returns on equity ranging from 20% to 35%, depending on how much leverage the project took on.

There is a range of opinion in the City of London and not every analyst was so sure that the deal was good for EDF. But our conclusion is that in the game of poker between the UK government and the French EDF, the British may have lost.

Conclusion

If HPC were to make very high profits for its investors, it would be embarrassing to the government in power when the station starts operating. This risk arises because the government deal inefficiently compensates investors for bearing the construction risk, which is temporary, through a very long-term price agreement. The government's gain share contract may provide assurance, but, as we shall see in Chapter 15, it initially fell short.

Meanwhile EDF and its partners still hadn't taken the "final investment decision" to sign contracts and start construction. One obstacle was the EC's clearance of state aid objections (Chapter 15). The other was the investment partners. EDF had described the Chinese companies as giving "serious expressions of interest". But just who were they and why were they interested in the British nuclear power market?

REFERENCES

Committee on Climate Change (2013). *Fourth Carbon Budget Review, Part 2. The cost-effective path to the 2050 target.* December. http://www.theccc. org.uk/wp-content/uploads/2013/12/1785a-CCC_AdviceRep_Singles_1.pdf

DECC (2010). *Consultation on a Methodology to Determine a Fixed Unit Price for Waste Disposal and Updated Cost Estimates for Nuclear Decommissioning, Waste Management and Waste Disposal.* March. https:// www.gov.uk/government/uploads/system/uploads/attachment_data/file/4 2533/1_20100324145948_e____ConsultationonFixedUnitPricemethod ologyandupdatedcostestimates.pdf

DECC (2015). *Contracts for Difference (CFD) Allocation Round One Outcome.* 26 February. https://www.gov.uk/government/uploads/system/ uploads/attachment_data/file/407059/Contracts_for_Difference_-_ Auction_Results_-_Official_Statistics.pdf

EDF Energy (2013). *Update on the UK Nuclear New Build Project ("NNB").* 13 October. http://shareholders-and-investors.edf.com/fichiers/ fckeditor/Commun/Finance/Publications/Annee/2013/EDF_NNB_ EquityPresentation_2_va.pdf

EDF (2015). *Annual Report 2014.* February. http://shareholders-and-investors.edf.com/fichiers/fckeditor/Commun/Finance/Publications/ Annee/2015/resultats_annuels/va/consolidated_financial_statements_ 2014_EDF_group.pdf

Energy Solutions (2015). http://www.energybrokers.co.uk/electricity/historic-price-data-graph.htm

European Commission (2014). *Decision of 08.10.2014 on the Aid Measure SA.34947 (2013/C) (ex 2013/N) Which the United Kingdom is Planning to Implement for Support to the Hinkley Point C Nuclear Power Station.* http://ec.europa.eu/competition/state_aid/cas es/251157/251157_1615983_2292_4.pdf

Guardian (2006). "Huge windfall for hospital's PFI investors as staff face job cuts". 3 May. http://www.theguardian.com/society/2006/may/03/ hospitals.privatefinance

House of Commons (2007). *Update on PFI Debt Refinancing and the PFI Equity Market.* Committee on Public Accounts. HC 158. 15 May. http://www.publications.parliament.uk/pa/cm200607/cmselect/ cmpubacc/158/158.pdf

Liberum Capital (2013). *Flabbergasted – The Hinkley Point contract.* 30 October.

NAO (2009). *Private Finance Projects.* October. http://www.nao.org.uk/wp-content/uploads/2009/11/HL_Private_Finance_Projects.pdf

NAO (2015). *UK Guarantees Scheme for Infrastructure.* HC 909. Session 2014-15. 28 January. http://www.nao.org.uk/wp-content/ uploads/2015/01/UK-Guarantees-scheme-for-infrastructure.pdf

Nuclear Energy Institute (2015). *Costs: Fuel, operation, waste disposal & life cycle.* http://www.nei.org/Knowledge-Center/Nuclear-Statistics/Costs-Fuel,-Operation,-Waste-Disposal-Life-Cycle

Taylor, S. (2016). *A Financial Analysis of the Hinkley Point C Project: Working paper.* http://www.jbs.cam.ac.uk/faculty-research/faculty-a-z/simon-taylor

14

Enter the Chinese
(2012-14)

A brief history of China's nuclear programme

Nuclear power's role in China's energy policy has been increasing since the mid-2000s, at which time it was still only a tiny fraction of the country's total electricity supply. China has abundant coal, which is cheap but which lies a long way from the main centres of population and industry, and is therefore costly to move. It is also highly polluting, annually causing thousands of early deaths and respiratory problems. It is also the main source of China's CO_2 emissions, the world's largest. China has expanded its use of gas, but most of it has to be imported. Nuclear offers a solution to all of these problems.

China's nuclear industry, like that of the UK, has its roots in the military. On 16 October 1964 China announced the detonation of its first nuclear device. China had the science and engineering skills to build civil nuclear power stations too, but this was not a priority for the next 30 years. It was only after Mao's death in 1976 that the Chinese economic miracle, with its extended period of 10% annual growth, started. From the early 1980s surging demand for electricity led to massive investment in power generation and the grid, but nuclear still wasn't a priority.

When China did decide to build nuclear power capability it recognized the difference between engineering nuclear weapons and building safe, economic nuclear power stations. China decided in 1970 to follow global mainstream technology and build a PWR. After a decade of research and development China developed its own version of the US PWR. A few years later a second project started, to bring the French version of the PWR to China.

China's first nuclear power station was a Chinese version of a

Figure 14.1: The Daya Bay nuclear power station

Source: Author

standard American Westinghouse PWR, which was constructed at Qinshan, in Zhejiang province, 100 km (62 miles) south-west of Shanghai. The CNP-300 started construction in 1985 and began operations in December 1991. Qinshan was built by and for the China National Nuclear Corporation (CNNC).

The second station was at Daya Bay, near Hong Kong. This was a French 944 MW PWR design built by the company Framatome (which later became part of Areva). Chinese engineers took part in the construction, which started in August 1987 and was managed by EDF, but the Chinese content was only 1%. It reached full commercial operation in early 1994 and was taken over by the company China Guangdong Nuclear (CGN), which was founded that year.

CNNC and CGN (which changed its name to China General Nuclear in 2014, reflecting its more global ambitions) are the two key Chinese nuclear companies that EDF intends to partner with on its UK nuclear project. Although both are state-owned enterprises (SOEs) they are somewhat different in character. CNNC is based in Beijing, geographically and politically close to the central government, and has been the major force behind the indigenous reactor designs. CGN is based in Shenzhen in Guangdong province, in the most commercial and historically outward-looking part of China. People who have dealt with both companies characterize CNNC as more long term in its focus, looking to the strategic national interest. CGN appears more commercial.

Both companies have links with foreign nuclear operators. CNNC is associated with Westinghouse, now owned by Toshiba, having licensed the PWR technology for the Qinshan reactor. Westinghouse's modern AP1000 design is now the main choice for the current expansion of China's nuclear fleet. CNNC is building two new AP1000s (another two are under construction for China Power Investment Corporation, one of the five big state-owned electric utilities). CNNC also has a long-standing relationship with EDF.

CGN has a similarly long relationship with the French nuclear industry, particularly with EDF. CGN licensed the French PWR design at Daya Bay to build its own indigenous and improved PWR, which it called the CPR1000. This Generation II PWR was the main reactor type used in the nuclear expansion in China until the government decided that all new reactors must be Generation III.

To build a Generation III (advanced safety) design CGN again teamed up with EDF and Areva. In February 2007 they agreed to build two of the new EPR (European pressurized-water) reactors at Taishan for a reported price of $8.5bn, much less than the cost of the equivalent plant being built at Olkiluoto in Finland. In November 2007 French President Sarkozy authorized a further two EPRs to be built. The EPR, the French contender in the global nuclear reactor competition, would therefore get a foothold in China, one of the largest new nuclear markets.

So the two Chinese nuclear companies had the money and the expertise and they knew EDF. The UK was an attractive market for them, both offering a large possible pipeline of future nuclear projects and representing a trophy investment in a major developed economy. The question was, how would the UK take to the idea of Chinese investment in a new nuclear power station?

The UK's China charm offensive

The coalition government of 2010 showed almost embarrassing enthusiasm for Chinese investment in the UK. Relations got off to a good start when Vice-Premier (and future Premier) Li Keqiang delivered a speech at a banquet in London in January 2011 that talked of expanded trade and cooperation on large projects (Government of China, 2011).

However, in May 2012 Prime Minister David Cameron "hurt the feelings of the Chinese people" by privately meeting the Dalai Lama at St Paul's Cathedral in London (BBC News, 2012). The Chinese government responded by blocking Cameron from visiting China. Once the British government made clear it got the point, contact between the two countries flourished. Chancellor of the Exchequer George Osborne visited China for five days in October 2013, including a visit to the EPRs under construction at Taishan. A large delegation accompanied the Prime Minister to China in December 2013.

Britain had two strategic needs with which China could help. The first was helping restore the City of London after the global financial crisis. The government was keen to make sure that the City reclaimed its leading role in international finance, which included making London the centre for trading the Chinese currency, the renminbi (RMB – "people's currency"), denominated in yuan. In January 2012 the UK Treasury declared that it wanted London to be the main hub for RMB trading outside Hong Kong, the only city where such trading was then allowed.

The RMB market was still very small owing to the restrictions on use of the RMB outside mainland China. But the Chinese central bank planned to "internationalize" the RMB by gradually removing barriers to the movement of funds into and out of China. So far official approval had only extended to Hong Kong, which, while a special administrative zone, was still part of China. London wanted official approval too.

In September 2011 Chancellor George Osborne announced jointly with the visiting Chinese Vice-Premier Wang Qishan their plan for London to become a hub for offshore RMB trading. The first RMB bond issued outside China and Hong Kong was launched in London in May 2012, by the bank HSBC, which is based in London but has deep historical roots in Hong Kong (HSBC originally stood for Hong Kong and Shanghai Banking Corporation). In June 2014, after the thaw in Sino-British relations, the UK and Chinese governments agreed that China Construction Bank, one of the big four state-owned Chinese banks, would be the first to act as official clearing agent for RMB trading in London. London's future as the main non-Asian trading hub for RMB currency and for RMB-related products now seems assured.

The British government's second interest in China is infrastructure investment. Funding the £466bn of projects in the 2014 version of the UK National Infrastructure Pipeline needs a lot of help. China has both funds and a lot of experience building infrastructure, including nuclear power stations. So it makes sense that Britain would want to encourage Chinese nuclear companies to invest.

The UK is generally relaxed about foreign ownership of its companies, including critical infrastructure. Parts of the gas, electricity, water, transport and telecoms industries are owned by investors from the USA, Canada, Australia, France, Germany and Singapore. Adding China is unlikely to upset the public, who seem to realize that the risk in infrastructure remains with the investors, who can hardly take their assets home if they're unhappy. There are some who worry about the prospect of a fully Chinese-owned nuclear station in future, because it would give China direct access to the national electricity transmission grid, perhaps the most critical piece of infrastructure of all. That prospect moved closer when the governments of the UK and China signed a "Statement of Cooperation in the Field of Civil Nuclear Energy" in October 2015, at the time of the state visit by President Xi Jinping. This "welcomed the proposal for a Chinese-led project at Bradwell", meaning the Chinese Generation III Hualong reactor which, subject to generic design approval, would be built in Essex, with construction starting in around 2021. Getting permission to build a wholly Chinese reactor in the UK would be a great boost to Chinese nuclear export prospects (and might provide cheaper power than the EPR). It might also mean they could do without their French partner EDF altogether.

Conclusion

EDF and the UK government are lucky to have the two Chinese companies, with their construction experience and finance. But that has not meant an easy or swift deal. At the end of 2015, the French, Chinese and British negotiators still hadn't reached the final investment decision for HPC.

REFERENCES

BBC News (2012). "David Cameron's Dalai Lama meeting sparks Chinese protest". 16 May. http://www.bbc.co.uk/news/uk-politics-18084223

Burr, W. and Richelson, J. T. (2001). "Whether to 'strangle the baby in the cradle': the United States and the Chinese nuclear program, 1960-64". *International Security* 25 (3), Winter 2000/01: 54-99.

Government of China (2011). *Chinese vice premier eyes intensified Sino-UK partnership.* http://www.gov.cn/misc/2011-01/12/content_1783208.htm

15

Getting state aid approval
(2013-14)

State aid as a bad thing

The EU bans most state support for private enterprise. This is to create a level playing field for business, the goal of the "internal market" of the EU. Ideally businesses should compete on their merits – on their ability to meet customer needs at the lowest cost. State intervention that rigs the game is both inefficient and unfair. (This had not always been EU policy, but the Lisbon Treaty of 2007 made it so and gave the EC the power to enforce it.)

Figure 15.1: EC's criteria for allowing state aid

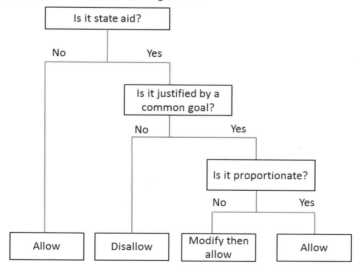

Source: author

Not all state aid is banned. There is provision for allowing aid under "exceptional circumstances" (Figure 15.1).

The UK government needed to convince the EC either that the Hinkley Point C (HPC) deal did not amount to state aid, or that the aid was justified by exceptional considerations that were compatible with wider EU policy goals.

The UK government had been keeping Brussels informed about its negotiations with EDF and officially notified the EC about the final deal on 22 October 2013. The EC published its first response on 18 December, suggesting profound scepticism about the legality of the government's proposals.

The EC at first appears opposed

The UK put its argument to the EC as follows: "failure to bring forward Hinkley Point C might translate into a complete lack of investment in new nuclear plants, as it might undermine the confidence of potential investors and industry about the feasibility of carrying out a project of such a financial scale" (European Commission (2013) 26).

According to the UK, HPC would not proceed without the three components of the deal with EDF: the CfD; the indemnity against future costs arising from political changes; and the sovereign credit guarantee. If there was no deal, there would be no new nuclear power station and therefore no decarbonization, or at least much less.

The proposed deal, in UK eyes, did not constitute aid according to the so-called Altmark criteria. (In 2003, in a case involving the company Altmark, the European Court had provided more explicit criteria for when state aid was acceptable or when certain state actions were not actually aid: the recipient must be carrying out public service obligations; the state could pay only the net costs of the service; the costs had to be clearly defined in advance; and ideally the payments should be determined through some form of public procurement process, but if not, they should be based on a strict examination of the true costs.)

The UK case was that the investment contract would not confer an advantage to the HPC holding company NNBG – rather, it was a condition for the investment to go ahead at all. NNBG would, through the generation of nuclear electricity, be contributing an economic public service. The EC concluded that both the CfD and the credit guarantee involved state aid.

As for the compensation for politically motivated shutdown, the UK argued this wasn't state aid either. But the EC needed to know whether it arose from a general principle and whether it would be available to other operators too. If it did, it would be non-discriminatory and would not unfairly help NNBG.

So the UK deal didn't appear to match the Altmark ruling's definition of state aid. The UK's next line of defence argued that if it was state aid, it was justified. The EC agreed this might be so if: i) the aid was necessary and proportionate; and ii) if the positive effects for the common objective outweighed the negative effects on competition and trade.

What objectives, common across the EU, might justify state aid? The first was decarbonization, which the EU had gone to great lengths to address, not least in setting up the Emissions Trading System (ETS). It was in part because the ETS carbon price had collapsed and so provided no financial incentive to invest in low-carbon energy that the UK had resorted to more direct intervention.

The EC agreed that decarbonization was an environmental goal but there was an environmental cost in the need to deal with radioactive waste: nuclear involved a trade-off. "The Commission therefore is not clear at this stage on whether the notified measure can be argued to be aimed at a common EU objective in terms of environmental protection in general, and decarbonisation in particular" (European Commission (2013) 246).

A second common objective was security of supply. The EC noted that the UK was pursuing a range of measures on security of supply. But given that the crunch point for generation capacity lay in the next five to ten years it was not clear that nuclear was essential for that goal, given the long construction time.

The final possible common objective was promoting nuclear energy itself, something of a stretch, given that Germany had recently decided to pull out of nuclear, and some other countries, such as Italy, had long since rejected nuclear power. The EC suggested that support for nuclear could be justified only on the basis that some form of market failure meant it was unfairly disadvantaged. A move to correct the market failure or compensate nuclear for that failure might be justified. But the EC was sceptical. "It is not clear that the current legal framework, or the characteristics of nuclear energy, result in a market failure. For the above reasons the Commission has doubts on whether the aid addresses a market failure related to electricity generation or to a specific market failure related to nuclear energy" (European Commission (2013) 297).

Overall the EC:

i) Questioned whether the state aid was necessary to achieve the stated goals of decarbonisation and security of supply (320);

ii) Doubted whether the instruments chosen (the CfD, credit guarantee and agreement to compensate NNBG) were appropriate, particularly if used together (the implication being that, combined, they amounted to too much help for NNBG) (333);

iii) Couldn't exclude the possibility that the credit guarantee would involve state aid, and that it might not be proportionate, risking overcompensation to NNBG (348).

In sum "if indeed aid exists, it would in principle be incompatible under EU state aid rules" (European Commission (2013) 8.1).

The government replied to the EC's preliminary thoughts on 31 January 2014. Publicly both the government and EDF expressed confidence that approval would be given; privately there were real concerns. The EC's opening thoughts had been more hostile than expected. The process of judging European state aid applications is a legal one, but it entails some ambiguity about what is legal, and politics are never completely absent. All state aid rulings risk pleasing or annoying other member states, which have their own agendas. In this case Germany was critical: not only did it now have a firmly anti-nuclear domestic policy, it was widely seen as the most powerful country in the EU and nobody wanted to annoy it unnecessarily. The EC's decision would have to be justified legally but there was concern about whether it might be influenced by German domestic politics.

Other parties sent comments to the EC, which then referred them to the UK, which returned its comments to the EC on 4 July 2014. There was then an agonizing three-month wait. To the delight of the government and the surprise of many commentators, on 8 October 2014 the EC approved a slightly modified version of the plan (European Commission, 2014).

October 2014: the EC pulls a rabbit out of a hat

Apart from confirmation that the project had been approved, the most striking news in the EC press release was that HPC was now expected to cost £24.5*bn*, compared with the previous £16bn. Since £16bn already made HPC the most expensive power station in the world, how on earth had the cost risen a further £8.5bn in just over a year?

The truth was it hadn't. In its own press release, welcoming the decision, EDF confirmed that the cost remained at £16bn in 2012 prices (EDF, 2014). The EC figure, which is not documented in the full EC report (or rather the figures are blocked out because they are commercially sensitive), included two adjustments. First, the £16bn figure was in 2012 prices so there was an inflation adjustment. This would be about 5%, raising the construction cost to £16.5bn.

The second adjustment was for contingent equity. Buried in the detail of the report was evidence that the government and/or the EC had required NNBG to make available £8bn of additional funding in case the construction cost overran. This would protect the government's guarantee on the debt, which would not be at risk of loss until that £8bn was used up. So in effect EDF and its partners had to guarantee they could fund construction even if it overran to £24.5bn. That was possible but unlikely, even given the EPR's previous overruns. So the new £24.5bn figure wasn't really an increase in the estimate of the construction cost, but included a provision for things going wrong. Even so, the £24.5bn figure dominated press coverage of HPC.

Why did the EC change its view? First, it took a newly sympathetic view of the measures as "investment aid", specifically intended to make it possible for NNBG to invest the "vast amounts of funds" necessary to build the HPC plant (European Commission (2013) 345) and now was convinced that there was indeed a genuine market failure with respect to nuclear power. "In particular, the promoters of the project would not be able to obtain the necessary financing due to its unprecedented nature and scale." State support for HPC should be allowed, subject to some modifications that shared with the electricity customer any gains from the project resulting from cost savings and NNBG beating its own targets.

The EC next argued that the CfD was purely a device for encouraging investment. It was not a form of procurement, as it promised to pay certain amounts in relation to electricity but wasn't specifically a contract to deliver that electricity. So the normal rules for procurement, in which several potential suppliers are invited to tender to make sure the state gets the best price, didn't apply. This was helpful to the UK since no other body was currently willing or able to tender for the CfD: the government had first published its plans for CfDs on 29 November 2012 (DECC, 2012) and NNBG had been the only nuclear operator with plans sufficiently advanced to respond. Formal negotiations between the government and NNBG started in February 2013. Since the process of identifying a suitable contractor for the CfD (ie a nuclear investor) had been clear and non-discriminatory, the EC found no grounds to question it. (This was a little surreal. It was fair enough to say that the government had not in any way favoured NNBG. But the process was non-discriminatory in the trivial sense that there was only one credible candidate.)

Next, the EC decided that the measures now did meet the common goal of energy security. It referred in particular to the Euratom Treaty of 1957, which it appeared to have forgotten about in its preliminary views. This treaty created the European Atomic Energy Community (Euratom), legally separate from the EU but governed by EU institutions. The EC noted that the treaty aims at creating "conditions necessary for the development of a powerful nuclear industry which will provide extensive energy sources". The treaty had been recognized in previous EC decisions, so it was appropriate for the current decision to acknowledge it; the UK could therefore invoke Euratom as a justification for promoting nuclear energy, and as part of a wider energy security goal.

The EC then explained why it now agreed with the UK that there was a market failure connected with nuclear energy that justified intervention. There was "merit" in the UK view that the absence of a long-term price of carbon justified state action to encourage low-carbon energy. There was a more general market failure in electricity markets, which might not supply enough electricity to keep the system stable and secure. The UK was proposing to fix this problem with a separate capacity market, in which the National Grid would pay extra funds to encourage the availability of power capacity when needed. While neither of these arguments is specific to nuclear, the EC argued there were two specific market failures that applied to nuclear and therefore justified state intervention:

i) Nuclear is subject to investment risk arising from the very high capital costs, long construction periods and very long periods of operation; the market doesn't provide risk management products that cover such long projects; the maximum length of contracts and of risk hedges is 10-15 years, but nuclear construction takes 8 years and the station is planned to operate for 60 years; therefore there is a market failure;

ii) Nuclear is particularly but not uniquely susceptible to what economists call "the hold-up problem" in which a company invests, irrecoverably, the capital to build a project and then a future government has an incentive to worsen the terms on which the product is purchased. This is in fact a generic problem with long-term capital projects in which the government is closely involved (ie most infrastructure investment); the EC noted that this wasn't strictly a market failure.

The EC argued that the *combination* of these two failures was unique to nuclear, which suffered corresponding difficulty in attracting private sector investment without state support. The EC had asked the DECC to compute various scenarios involving fossil fuel prices, with the new capacity market in place and with a carbon price floor, another part of the government's policy mix to encourage low-carbon generation investment. In none of these scenarios did nuclear appear to make private sector financial sense until the early 2030s at best. The EC agreed that without CfDs there was a high risk that no new nuclear investment would be forthcoming for decades at best.

Does nuclear suffer from market failures?

The EC's argument was somewhat doubtful from an economic, if not necessarily a legal point of view. A market failure is a well-defined concept in microeconomics. Theory suggests that under idealized conditions, a market delivers efficient outcomes: demand and supply are balanced, costs are minimized and consumer wants are met as far as possible. Under these conditions there is no case for governments to intervene because they are likely to spoil things. But the idealized conditions often fail to hold in reality. A market failure is when one of those conditions is absent and there is a case for state intervention to either correct it or take actions that offset it or minimize its impact.

The EC argued that it agreed with the UK, that no nuclear investment would happen without some package of incentives and the UK package was duly proportionate to that need. But that is not a market failure. There are many private sector investments that don't proceed because investors decide the return is not high enough to compensate for the risk. A large nuclear power station is one example. But where is the market failure?

The true market failure is that carbon emissions are not "priced" in the electricity market. The ideal solution to that would be a market for emissions that sets a price that correctly reflects the social cost of emission. The European ETS was intended to do precisely that, albeit only for the EU and with quite significant omissions and limitations. If the ETS worked well then private investors would be incentivized to build low-carbon energy generation, perhaps including nuclear if it stacked up against other low-carbon sources such as wind or solar.

But for the government to decide that nuclear was essential for its policy and then to argue it wouldn't happen without state help was a more tortuous argument. In effect the UK and the EC were arguing that the end justified the means. Given the urgency of combating climate change and given the realities of the state of the nuclear industry (a small number of large companies with reactors that needed huge investments, which their shareholders were unwilling to fund without some form of government risk sharing) the argument was that nuclear was necessary, and therefore state aid was necessary and justified because otherwise it wouldn't happen.

Opponents argue that nuclear is only one form of low-carbon power and should be made to fight it out, in a non-discriminatory way, with other low-carbon sources. That was indeed the earlier UK government policy, but the ambitious decarbonization targets, combined with the imminent closure of older nuclear stations and the difficulty of increasing the scale of onshore wind, left the government with little choice but to do whatever it took to bring private nuclear investment to the UK. The EC's explanation of why it now agreed with the UK government was therefore not entirely convincing.

The extra conditions

Although it accepted the principle of state aid, the EC wanted more to be done to make sure that the intervention measures did the least damage to the market and to the interests of consumers who would ultimately pay for them: in EC-speak they needed to improve the "proportionality" of the measures. They had therefore negotiated a combination of two modifications to the original package that "minimises the distortive effects of the support measure and ensures benefits to UK consumers".

First, NNBG would pay a higher fee for the state credit guarantee, which would cover up to £17bn of debt. The original fee was to have been 2.5%. But it would now be 2.95%. This would raise the amount paid to the government by £1bn. It was an essential part of the UK case that the guarantee would be charged at a commercial rate, there would therefore be no state subsidy. But the EC believed 2.5% was below market rates, which implied there would have been a subsidy.

If this had been a standard form of financial insurance, then it would have been possible to check the going rate in the market and charge that. But the whole point of the government credit guarantee was to offer something that could not be found in the commercial market. Only a government could offer a guarantee for up to £17bn of debt, so large was the potential risk. The government credit guarantee was a unique product for which private sector benchmarks would be virtually impossible to find. The EC had to estimate somehow a "hypothetical market rate for a facility which is not offered by the market" (European Commission, 2014: 475). On the basis of comparisons of the credit costs of a range of projects in the infrastructure field the EC determined that the fee should be 2.95%.

Second, the EC wanted to make sure that there was a fair mechanism for sharing gains between the investor and government (or ultimately the electricity customer). The gains could arise from two sources: i) savings on construction costs; and ii) a higher than expected rate of return to investors. The first was particularly important, given that the DECC had estimated a range of construction costs from £10bn to £18bn with a median case of £13bn. This amounted to a very wide range of outcomes, from investors having to put a lot more funding into the project (the so-called "contingent equity") to finding themselves making a saving, with the swing being billions of pounds.

The exact details of the construction cost "gain share" are not public. The EC decision reads:

> The new construction gain share will provide that:
> a. The first GBP [...] billion of construction gain (nominal value) will be shared on a 50:50 basis with 50 per cent of the gain going to the CfD Counterparty and 50 per cent to NNBG; and
> b. Any construction gain in excess of GBP [...] billion (nominal value) will be shared on a 75:25 basis with 75 per cent of the gain going to the CfD Counterparty and 25 per cent to NNBG.
>
> (European Commission (2014) 487)

The dots [...] show figures kept out of the public domain on grounds of commercial sensitivity. But there is at least a mechanism, where before there was none.

On the rate-of-return gain share, the EU strengthened the deal in favour of the electricity customer. The 10% post-tax return figure that had been widely discussed was the overall return – on equity and borrowed money combined. The higher (2.95%) credit guarantee cost would bring this return down to about 9.5%. According to the

DECC and checked by the EC, this was equivalent to an *equity* return of 11.25% under the more stringent EC-mandated sharing scheme (compared to 12.5% in the UK's initial August 2014 gain share proposal).

The new EC gain share formula was:

a. Any gain above and beyond [11.4%] will be shared by the CfD Counterparty for 30 per cent and by NNBG for 70 per cent.
b. A second threshold set at the higher between 13.5 per cent in nominal terms or 11.5 in real (CPI-deflated) terms, based on the same model as in point a above. Above this threshold, any gain will be shared by the CfD Counterparty for 60 per cent and by NNBG for 40 per cent.

(European Commission (2014) 488)

This equity gain share mechanism would be in place for the entire 60-year lifetime of HPC, not just the 35 years of the CfD. It kept the majority of the gain in returns for the investors to encourage them to keep costs down. But the gain share gave some reassurance to the public that the project couldn't generate unexpectedly high returns for the investors without some reasonable share of those gains being given back to them.

The EC's amendments would limit the profit potential for EDF and its investment partners but it would also quell public concerns that, if the project went well, particularly if this were the first EPR to actually be built on time and budget, then the investors would make outrageously high returns on the back of electricity prices that might seem astronomically high in 20 years, though nobody could know that now. Although the EC had delayed the process and published more about the commercial details than the government and EDF might have liked, the EC had improved the acceptability of what could otherwise be an exceptionally controversial project.

Conclusion

The price of securing the HPC investment was a long-term price promise, a credit guarantee and indemnity against future policy changes. But the EC conditions made it less likely that, in the game of poker between EDF and the government, EDF would walk off with all the chips.

EDF and its partners had everything they needed: a package of support from the government; approval for the reactor; and planning permission. So why didn't they just go ahead?

REFERENCES

DECC (2012). *Annex A: Feed-in Tariff with Contracts for Difference. Operational framework.* 29 November. https://www.gov.uk/government/uploads/system/uploads/attachment_data/file/66554/7077-electricity-market-reform-annex-a.pdf

EDF (2014). *Press Release: Agreements for planned Hinkley Point C nuclear power station approved by the European Commission.* 8 October. http://press.edf.com/fichiers/fckeditor/Commun/Presse/Communiques/EDF/2014/cp_EDF_20141008_va.pdf

EU (2008). *Treaty on the Functioning of the European Union.* http://eur-lex.europa.eu/legal-content/EN/ALL/?uri=CELEX:12008E108

European Commission (2013). *State Aid SA. 34947 (2013/C) (ex 2013/N) – United Kingdom investment contract (early Contract for Difference) for the Hinkley Point C new nuclear power station.* http://ec.europa.eu/competition/state_aid/cases/251157/251157_1507977_35_2.pdf)

European Commission (2014). *Decision of 08.10.2014 on theAaid Measure SA.34947 (2013/C) (ex 2013/N) Which the United Kingdom is Planning to Implement for Support to the Hinkley Point C Nuclear Power Station.* http://ec.europa.eu/competition/state_aid/cases/251157/251157_1615983_2292_4.pdf

PART V

Conclusion

16

The UK nuclear power outlook in 2015

Hinkley Point C: limping towards the finishing line

After the European Commission (EC) approved the UK's investment support for HPC in October 2014, it seemed that at last EDF had everything it needed to take the "final investment decision" (FID) that it had been promising for at least three years. But still nothing was announced.

The underlying financial difficulties in the French nuclear industry were becoming more apparent. On 21 October investment bank J. P. Morgan forecast that for the rest of the decade, EDF's net cash flow (after paying dividends) would be less than its capital spending, mainly because of the ever-increasing cost of refurbishing its French reactors and preparing them for life extensions (J. P. Morgan, 2014).

And in November EDF's partner, Areva told investors it could no longer confidently give a profit outlook because of the uncertain cost of its involvement in the Finnish Olkiluoto nuclear power station fiasco. Areva was supposed to be contributing 10% of the investment in HPC, a commitment that was now looking doubtful. Concern at the UK Treasury was enough to instigate a review of whether the project could survive without Areva's funding.

Rumours over the HPC shareholder structure continued – state-owned Saudi Electric was said by the *Financial Times* to be considering taking a 10% stake. Kuwait and Qatar were also mentioned, suggesting that anybody with access to several billion pounds would now be welcome.

A few days later the UK political support for HPC showed its first crack when Labour's Shadow Energy Minister, Tom Greatrex, urged the National Audit Office (NAO) to investigate whether HPC still represented value for money. This was a very reasonable question, but the NAO, while reportedly very keen to look at the contracts, was worried about investigating before the project was finalized for fear of jeopardizing it (*Financial Times*, 2014).

On 4 December EDF Energy's indefatigable boss, Vincent de Rivaz made his latest forecast, saying he was confident of a decision by the spring of 2015.

On almost the same day the Finnish parliament demonstrated they had learned the lessons from the disastrous overrun at the Olkiluoto EPR by approving another new nuclear power station. This one would be supplied by the Russian company Rosatom. The latest estimate for Olkiluoto was that it would open in 2018, 13 years after it started construction. A Finnish newspaper had decided early in the year that it was probably the most expensive building in the world, likely to cost €8.5bn (*Helsingen Sanomat*, 2014).

By contrast the consortium of companies buying the Rosatom reactor were confident it would be built on time and on budget and would produce power at €50/MWh (£36/MWh, compared with the £92.50 agreed for HPC) (World Nuclear Association, 2015). With two Russian reactors already operating satisfactorily (with Western-designed containment vessels) the Finns were happy to go for the lowest price.

Vincent de Rivaz's forecast turned out to be as accurate as his many previous ones. On 12 February 2015 he announced that the decision wouldn't be taken in the first quarter after all. There were further discussions about the government contract. It was reported that EDF was seeking a government indemnity against the risk of a legal challenge to the EC state aid approval from Austria, which had earlier threatened to challenge the ruling. While the government was confident that any challenge would be defeated, it posed a small but unwelcome risk.

Back in France, Areva's financial problems were mounting. In March the company announced a €4.8bn (£3.5bn) loss, greater than the company's market value of €3.6bn. The company's main shareholder, the French government, would need to put in new equity to stop Areva going bankrupt. Its debt had been downgraded in 2014 by rating agency Standard & Poor's to "below investment grade" (more commonly called "junk"). It was now extremely unlikely that Areva would be able to fund its intended 10% stake in HPC. The most obvious (to the French government) solution was for EDF to buy or merge with Areva, but that was not at all attractive to EDF, which was already much weaker financially than it had ever been. Vincent de Rivaz affected a relaxed attitude and told reporters that Areva's difficulties were "not existential for Hinkley Point" (Bloomberg, 2015). The new Areva Chief Executive, Philippe Knoche, told the *Financial Times* that "Our commitment to Hinkley cannot be questioned in any manner", just the sort of statement that invites the observation that he doth protest too much (*Financial Times*, 2015).

Although the government and EDF appeared to have ironed out their main contractual issues at the time of the announcement of the financial package in October 2013, there were still hundreds of points to be resolved. The process was slow and legalistic, and dragged on into the second half of 2015.

Meanwhile the parallel negotiations between EDF and the Chinese were proving fractious. Although EDF had worked with both China National Nuclear Corporation (CNNC) and China General Nulear (CGN), their interests were increasingly divergent. The Chinese, particularly CGN, were seeking a leading role in the world nuclear industry. They were not happy to be just a source of finance for a project that would be dominated by HPC. And they wanted assurances from the government that they would have a leading role in a future nuclear project in the UK at Bradwell in Essex, next to the now closed Magnox station.

The British government had shown every sign of wanting to encourage Chinese investment in the UK and the Chinese could plausibly claim to have the world's greatest expertise in nuclear construction. But they also wanted to bring their own reactor technology to the UK, which would require independent approval by the Office for Nuclear Regulation (ONR), something that politicians could not promise would be given.

EDF had put around £2bn into HPC and had the prospect of a very reliable long-term profit stream to look forward to, if they could build the EPR on budget. The Chinese, while certainly not looking to throw money away, had a possibly unique opportunity to enter a major developed country market, opening up the possibility of a wider international presence in future, with or more likely without EDF. The government was worried that if HPC were delayed further or even cancelled, then it would face a very large gap in its plans for future decarbonization.

Better to cancel HPC?

The betting was that a deal would be done. Yet there was a mounting argument for delay. First, HPC had ended up with an inefficient risk allocation that would probably give excessive returns to the investors in exchange for their bearing all of the construction risk. If the government were willing to take some construction risk, it could negotiate a lower long-term power price deal. Second, the other nuclear projects, which offered a good (but not guaranteed) prospect of lower costs because they used different reactor types, were making progress. Allowing for likely shorter construction times, it was quite possible that the Moorside project near Sellafield in Cumbria, using the Westinghouse/Toshiba AP1000 reactors, might come on stream ahead of HPC, even with construction unlikely to start until 2020.

Another reason for delaying or scrapping HPC was to wait and see how other energy technologies developed, what economists call the "option value" of waiting. The first competitive auctions for CfDs for renewable energy in February 2015 had shown encouraging falls in the prices investors required, with offshore wind coming in at £120/MWh and onshore wind at £82.5/MWh. Proponents of solar PV were increasingly confident that the dramatic fall in prices seen in the last 10 years would continue, to the point where, within a decade, it might become economic without subsidy (see below). Perhaps the vast and inflexible investments in nuclear might not be needed after all, or at least not on the same scale as previously thought.

There were also two counter-arguments. The first was that nuclear was central to any realistic achievement of the government's decarbonization targets. The various scenarios put forward by the CCC from 2009 had envisaged a large role for nuclear, tempered in part by the use of carbon capture and storage (CCS) as an alternative. The problem was that CCS was at least five years behind schedule and not yet proven to work. In the words of the CCC's 2015 progress report to parliament "In our first report in 2008 we set out that CCS urgently needs to be proven and commercialised in the UK, and our original indicators reflected this with the first plant coming online in 2014 and multiple plants by 2020. Progress has been slow with the first projects now aiming to commence operation by 2020" (CCC (2015) p. 57). While nuclear was also six years behind the schedule envisaged by the CCC in its 2009 report, it was at least known to work.

The upshot was the CCC was now worried about the government hitting its fourth carbon budget, for 2023-27, which had been set in June 2011, following advice from the CCC in December 2010, at 390 $MtCO_2e$ (compared with 566 $MtCO_2e$ in 2011). In its July 2014 progress report the CCC warned that "our assessment is that there will be insufficient emissions reduction under current policies, both in power generation and energy-intensive industries" (CCC, 2014: 29). In July 2015 the CCC found that "a similar amount of current ambition [announced policy actions] remains at risk" (CCC (2015) 29).

The CCC's job is to report to parliament on progress and warn if a target is in danger of being missed. What happens next is unclear; the Secretary of State would be in breach of a legal obligation so a legal process might ensue to force compliance. Alternatively, the law could be abolished if parliament agreed.

So new nuclear was increasingly necessary for hitting the legal carbon targets. Despite the delays and increased costs, the government, not wanting to face the public criticism of the CCC and a possible legal challenge, needed to keep going with whatever low-carbon options it had, including HPC.

The second argument against scrapping HPC was that this project, the most advanced and closest to completion of the various nuclear projects in the pipeline, could cause a loss of momentum and confidence that might sabotage the entire new nuclear programme. That was certainly the view of many civil servants and politicians. The best (waiting for a cheaper nuclear station based on either ABWR or AP1000 reactors) might therefore be the enemy of the good (an expensive but workable station that would start construction in 2016, not 2020).

Other projects on track

Other new nuclear projects continued to make progress (Figure 16.1).

Figure 16.1 represents a feasible but probably optimistic schedule. The stations most likely to make rapid progress are the ABWRs (advanced boiling-water reactors) planned by the Hitachi-owned Horizon company. The ABWR was certified by the US regulator in 1997, has a good record of being built on time and on budget, and Hitachi (with its partner the US company GE) is strongly committed to it. Although the UK has never had a boiling-water reactor before, this is a standard design that is used in the USA and Japan, and it should receive generic design approval in 2018. With a construction time of less than six years it is therefore realistic that it could be

Figure 16.1: New nuclear power station projects in the UK

Station	Sponsor	Reactor type	Gross GW capacity	Projected commissioning
Hinkley Point C	EDF	EPR	3.3	First unit 2024
Bradwell B	EDF	EPR	1.7	First unit 2030
Moorside	NuGen	AP1000	3.4	First unit 2026
Oldbury C	Horizon	ABWR	1.6	First unit 2028
Oldbury-on-Severn	Horizon	ABWR	1.6	First unit 2030
Sizewell C	EDF	EPR	3.3	First unit 2026
Wylfa Newydd	Horizon	ABWR	2.8	First unit 2024

Sources: National Grid 10-Year Statement; company websites; author's estimates

operating by 2024, quite possibly before HPC is commissioned.

EDF also has a project at Bradwell in Essex, which is now earmarked for a Chinese-led reactor, pending generic design approval likely to start in 2016. Meanwhile the other competing reactor design, the Westinghouse AP1000, is intended for a project at Moorside in Cumbria. The AP1000 is a more revolutionary design than the ABWR, but is under construction in the USA and China, with some reported delays, though less serious than those of the EPR. The AP1000 raised more issues for the Office for Nuclear Development (OND), and the generic design approval (GDA) process was temporarily suspended, but restarted in 2014. Westinghouse (owned by Toshiba) hopes for GDA by 2017 and plans to be online by the mid-2020s.

Much could go wrong with these plans, not least the financing. Although not quite as expensive as the EPR, the other reactor designs will still require about £10bn, allowing for the same contingent equity as HPC. That is a large amount even for companies the size of Hitachi and Toshiba, and it remains to be seen how they can raise it.

The other constraint is engineering and construction. The UK's relatively small pool of engineers experienced in major project construction will be stretched building one nuclear station, let alone three or four. Things will be even tighter if the High Speed Rail 2 (HS2) project goes ahead in the mid-2020s. There are also bottlenecks in the manufacturing of key components for nuclear power stations, which require steel forging on a scale that only a handful of companies can undertake. Worldwide nuclear reactor construction could therefore become congested quite easily.

Existing stations

The best news for the CCC is that the existing British nuclear stations are likely to produce power for several more years than expected, thanks to the life extensions of the advanced gas-cooled reactors (AGRs). Figure 16.2 shows an estimated outlook for the eight existing British stations.

Figure 16.2: Projected life of current UK nuclear power stations

Station	Type	Gross capacity (GW)	Projected life
Dungeness B	AGR	1.1	Extension to 2028 agreed Jan 2015
Hartlepool	AGR	1.2	Five-year extension granted from 2019
Heysham 1	AGR	1.2	Five-year extension granted from 2019
Heysham 2	AGR	1.2	Five-year extension granted from existing 2023
Hinkley Point B	AGR	0.9	Extension to 2023 granted 2012
Hunterston	AGR	0.9	Extension to 2023 granted 2012
Sizewell B	PWR	1.2	Continues to at least 2035
Torness	AGR	1.2	Five-year extension from existing 2023
Wylfa	Magnox	0.5	Assume closes end 2015

Sources: National Grid; EDF Energy; author's estimates

Several AGRs have received regulatory permission for five extra years beyond their current permitted life. Dungeness B, the station with the worst operating and construction history, received permission in January 2015, eliciting a wry smile from those familiar with "the single most disastrous engineering project undertaken in Britain" (see Chapter 2).

It's plausible that all the extensions will receive permission, which is what Figure 16.2 assumes. Some may even be extended further. A life extension brings two financial benefits: i) five more

Figure 16.3: Projected UK nuclear generation capacity (GW)

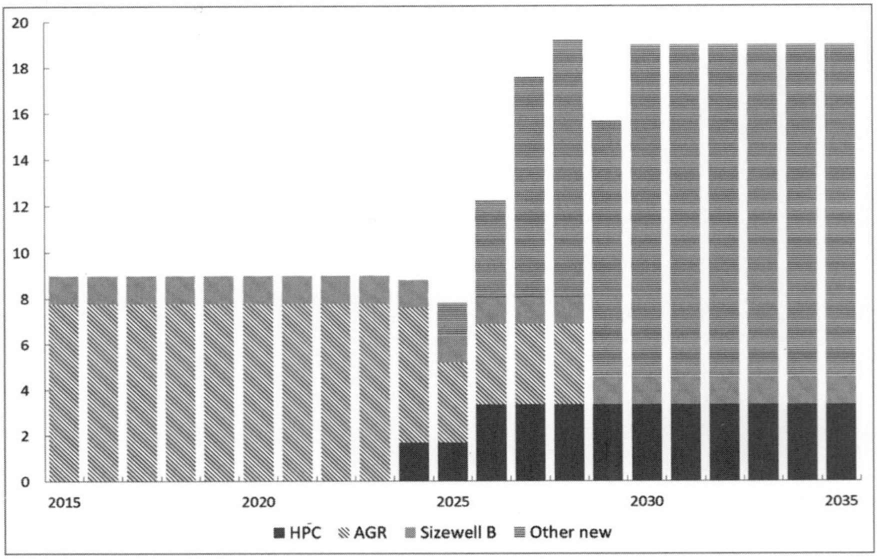

Source: Author's estimates

years of cash generation; and ii) five years' delay in the future cost of decommissioning. Each extension is both a regulatory matter and an economic one. In some cases the cost of upgrading the station may not be worth the benefit of five years of extra generation, even if the regulator allows it.

The Sizewell B PWR is projected to continue to at least its 40-year design life of 2035 and likely to get extensions beyond that.

Putting all of these tentative assumptions and projections together we get a projected path of UK nuclear capacity as shown in Figure 16.3.

Figure 16.3 makes fairly optimistic assumptions about the existing stations and the progress of new ones. Another setback for HPC could delay or even kill the other projects. In our optimistic scenario the UK has a temporary fall in nuclear output in 2025, but then new stations come on stream just as the AGRs are gradually retired, leaving about 19 GW of capacity operating from 2030 (about 15-20% of total generation capacity).

Figure 16.3 projects additional new nuclear capacity of about 17 GW by 2030. Sizewell B will contribute a further 1.2 GW. Is that enough to hit the fourth carbon budget? The CCC suggested 12-17 GW of new nuclear is needed by 2030, so this would hit that target. But the overall budget requires a package of 24-40 GW of offshore wind, up to 10 GW of CCS (the most speculative assumption given CCS's lack of progress so far) and 13-25 GW of onshore wind. It is doubtful that all of these other low-carbon targets will be hit, given the rising hostility to onshore wind in particular.

Indeed none of the low-carbon options looks easy or assured. The CCC, in its 2015 review of progress on emission targets, sounded understandably cautious:

> In order for a successful programme of new nuclear plant to be deployed, projects need to deliver to time and budget. If costs rise and the benefits of a programme do not translate into lower costs than for the first plant (i.e. £92.50/MWh, which has been agreed for Hinkley Point C) then the value of a nuclear programme could be called into question, particularly if other low-carbon options are making good progress.
>
> (CCC (2015) 56)

However, it is equally likely that CCS will fall short, putting more pressure on nuclear. The first "at-scale" CCS demonstration plant started operating in Canada in 2014, but power costs for early plants have been projected at £150-200/MWh, higher even than nuclear (CCC (2015) 57).

One area of hope is solar PV, the cost of which has been falling and is widely expected to continue to do so. Nuclear is behind schedule, CCS remains unproven and onshore wind is unpopular. But the CCC reported that solar PV costs had fallen to £122/MWh from its 2011 estimate of £315-460/MWh. DECC estimates that further cost falls may make solar competitive with onshore wind and nuclear power by 2020, and cheaper than offshore wind (CCC (2014)118).

But critics point out that the intermittency of renewables requires backup supplies that might have to be gas-fired, damaging both the economic and environmental case.

That raises the stakes for new nuclear. It is unlikely that solar could be deployed fast enough to avoid the need for some new nuclear power investment (recall David MacKay's "all south-facing roofs" projection, Chapter 7). But it strengthens the case for waiting a bit longer to see. As well as cheaper solar, future nuclear options such as small modular reactors (SMRs) offer the possibility of cheaper and more flexible smaller-scale nuclear deployment, reducing both the large-project risk and the unbalancing of the national transmission grid due to a few very large stations creating high localized concentrations of capacity – much higher than the grid was ever constructed for. SMRs are making progress in the USA, which is facing a huge retirement of its older nuclear fleet in the next 20 years. SMRs may turn out to be yet another of nuclear innovation's failed hopes. Alternatively, they may offer a way, for the first time, to achieve economies of scale by using standardized designs that allow components to be manufactured in a production line instead of as one-off and ad hoc design projects.

Conclusion

In April 2015 EDF and Areva told the French nuclear regulator of "manufacturing anomalies" in steel components particularly important for safety at the Flamanville EPR construction site (BBC News, 2015). Further tests would follow. These anomalies could further delay HPC and potentially even stop the EPR completely. In June, the French government confirmed that EDF would buy the reactor business of its state-owned partner Areva, ruined by its liability for cost overruns at the Olkiluoto EPR. The burden on EDF would raise further doubts as to its ability to finance its share of HPC.

By the time that China's President Xi Jinping arrived for his state visit in October 2015, expectations had built up that the Hinkley deal would finally be signed. A deal was indeed done, but only for non-binding heads of agreement. This was progress, but not yet the final investment decision that would lead to construction starting. The Chinese and UK governments committed themselves to the project, making it all but certain that it would go ahead. But paradoxically the state commitment allowed each side's lawyers to play hard in continuing negotiations, each betting that the other wouldn't allow the project to be delayed further.

The regulators were also reported to be asking for modifications to the troubled EPR. EDF quietly raised their estimated construction cost from £16bn (in 2012 money - not very different from 2015 money with the UK's low inflation) to £18bn. EDF would now take two thirds of the project, the other third would come from CGN, one of the two Chinese nuclear companies. EDF hoped to sell down to 50% later but it was not clear where the other investment was going to come from. Despite optimistic noises from EDF about reaching final investment decision within weeks, it seemed unlikely that the project would actually start until well into 2016.

The Chinese got what they really wanted, a British government promise (subject to regulatory approval) that they could lead a project at Bradwell in Essex using their own fully Chinese Hualong reactor

With HPC delayed, CCS far from proven and rising hostility to onshore wind, the prospects of the UK hitting its fourth carbon budget looked discouraging in 2015. But nuclear's proven ability to replace large amounts of CO_2 keeps it central to policy. The Climate Change Act was intended to force governments to make steady and accelerating progress towards decarbonizing the British economy. That risks turning into pressure to get new nuclear investment, at almost any cost.

REFERENCES

BBC News (2015). "Nuclear power: UK 'must learn' from French reactor concerns". 18 April. http://www.bbc.co.uk/news/uk-32365888

Bloomberg (2015). "EDF sees U.K. backing for nuclear no matter who wins election". 5 March. http://www.bloomberg.com/news/articles/2015-03-05/edf-sees-u-k-backing-for-nuclear-no-matter-who-wins-election

http://www.carbonbrief.org/blog/2014/11/how-the-uks-nuclear-new-build-plans-keep-getting-delayed

CCC (2014). *Meeting Carbon Budgets – 2014 progress report to parliament.* July. http://www.theccc.org.uk/wp-content/uploads/2014/07/CCC-Progress-Report-2014_web_2.pdf

CCC (2015). *Meeting Carbon Budgets – Progress in reducing the UK's emissions. 2015 report to parliament.* 30 June. http://www.theccc.org.uk/wp-content/uploads/2015/06/6.737_CCC-BOOK_WEB_250615_RFS.pdf

Financial Times (2014). "Labour seeks probe of Hinkley Point C nuclear project". 20 November.

Financial Times (2015). "Areva's new boss promises shift in strategy". 15 March. http://www.ft.com/cms/s/0/51265628-c978-11e4-a2d9-00144feab7de.html#axzz3VypdeeyC

Helsingen Sanomat (2014). "The new Olkiluoto power plant is already more expensive than any skyscraper" (translated from Finnish). 2 April. http://www.hs.fi/talous/a1396324226210J. P. Morgan (2014). *EDF – New tariff framework suggests unappealing prospects for minority shareholders.* 21 October.

World Nuclear Association (2015). *Nuclear Power in Finland.* http://www.world-nuclear.org/info/Country-Profiles/Countries-A-F/Finland/

17 / Conclusion

In 1979 Stephen Pile wrote the best-selling *Book of Heroic Failures: Official handbook of the not terribly good club of Great Britain*. It appealed to the self-deprecating side of the British people at a point in their history when the long post-war recovery seemed not to have gone as well as it should.

Nuclear power in the UK has something of the air of heroic failure. There are no villains in the story of how the UK went from being a pioneer in an exciting technology with real hopes of world success, to having no reactor design of its own and having to attract French, Japanese and Chinese expertise and money. There are plenty of individual successes, fine intentions and grand visions. Those visions appeared to have vanished at the start of the 2000s, when nuclear power appeared hopelessly uneconomic and ill-suited to the competitive power market.

The UK had created that competitive environment as part of its long flirtation with market forces. The Thatcher government's idea was that energy policy could be left to the market. It took years for civil servants to accept this, but they did, and subsequently they and their successors became resistant to the idea that there was any role for the state. However, the new ideal of a climate change policy amounting only to corrective taxes or a price on carbon, with the market otherwise left to make the right choices, collided with the Climate Change Act reality of five-year carbon budgets and central planning.

Keeping the state's financial exposure to a minimum (as embodied in the "no subsidy" mantra) pushed the risk on to private investors, who would only accept it with a 35-year power price contract and a state debt guarantee. In thus avoiding any exposure to the construction cost risk of a new nuclear power station, the government saddled a generation of electricity customers with potentially very expensive power prices. The likelihood that a second generation may later get very cheap power from a fully depreciated nuclear station was not much comfort.

A gradual, and presumably unintended shift can be seen in governmental language – from "allowing" the private sector to invest in nuclear in 2008 to "urgent need" in the DECC's *National Policy Statement for Nuclear Power* (DECC (2011) 2.5.2). But the language of no public subsidy continued – politicians were not being entirely honest with the public.

The fact that in 2016 there is a real possibility that a new nuclear power project, Hinkley Point C (HPC), might actually start is, for some, a heroic achievement. A decade earlier such a project seemed impossible. The amount of public policy work that went into reforming the planning system, restructuring the electricity market, reorganizing nuclear regulation and reactor approval, a vast array of energy white papers, reviews and discussion papers, is remarkable. The extent of public consultation is impressive. Nuclear has gone from being the most secretive, furtive and distrusted part of the public sector to an emblem of open government, in which every decision is carefully described, explained and justified before being put out for discussion. The documents are very clearly written, the underlying assumptions are documented and the data are revealed.

So it is with some pride that civil servants point to the fact that the UK is close to building a 3.2 GW low-carbon power station that could be operating until the end of the 21st century. And there are one, maybe two or three more projects following behind. HPC is not an aberration; it is the first, a little tardy admittedly, of several projects to come.

The critics fall into two opposing camps. The first says: why on earth has it taken so long? Why is the UK able to pass a policy in favour of new nuclear power in 2008 and eight years later still have nothing more to show for it than preliminary earthworks and displaced badgers? Why can't the UK deliver major infrastructure on time? The private sector is willing and able to mobilize billions of pounds if only the government would set clear rules and stick to them.

The second camp says: why are we (and it is we, every user of electricity in the UK) spending such extraordinary amounts of money on machines that might seem modern but are rooted in a mid-20th-century view of energy? HPC will be a cathedral to the cost of the baroque engineering needed to make the pressurized-water reactor even safer than it already is. Wait just a few years, this critic says, and we could be on the cusp of a solar revolution providing affordable power, decentralized across the nation and avoiding the risk of grid breakdowns stemming from the sudden failure of giants like HPC.

One day we may know which, if either, of these critics is right. Politicians and civil servants must take decisions now, with imperfect knowledge of the future. The private sector was very confident in 2007 that it could build new nuclear without subsidy. That same private sector a few years later, wounded by a financial crisis and finding that construction of the EPR was going very badly, had to ask for first a carbon price guarantee, then a long-term power price contract and a debt guarantee. Even that wasn't enough. It required additional investors, the terms of whose involvement has not been an easy matter to settle.

Civil servants have done their best to reconcile government targets with market forces but they lack experience in major projects, they have too little relevant private sector finance advice and they appear still prone to opt for the best rather than something merely workable. Backing a new, unproven reactor design (the EPR) looks rather like repeating the mistakes of the 1960s. But cancelling it now, despite the possible savings, risks jeopardizing the entire nuclear programme and missing the carbon budget.

The government felt it must choose the super-safe reaction option, whatever the extra cost, over the merely very safe. Such are nuclear economic choices, at least until cheaper modular designs are proven, perhaps after 2020.

If Hinkley goes ahead, it will open the way for at least three other, probably less costly and quicker, nuclear projects. If financial or technical obstacles, or the inability of investors to agree terms, lead to the cancellation of Hinkley, the other nuclear projects would become critical to meeting the UK carbon plans. With older nuclear stations heading for closure, the UK would find it extremely difficult to decarbonize the electricity supply by 2030 without some new nuclear. This book has shown how government policy has made nuclear logically central to the UK's climate change policy.

Alternatively, a future government, facing higher power prices and little evidence that other countries were following the UK's lead, might scrap that policy and aim for mitigation rather than prevention of serious environmental costs. The momentum in 2015, with agreements between the USA and China, growing use of carbon markets in the USA, Canada and China and a gradual shift in public opinion in the USA, all pointed to an increasing likelihood of serious action on climate change. Against this backdrop the UK is less likely to abandon its policy. And energy security concerns continue to make a separate case for new nuclear investment.

Nuclear is at the extreme end of the spectrum of infrastructure investment. Infrastructure is essential to the economy but highly problematic for a private investor. The very long timescales, dealing with politicians who change their views in response to a fickle public and media, and the risk that future governments will fail to honour the promises of their predecessors – these are all impediments to the commitment by pension funds and life insurance companies of their customers' money; these customers being the mass of ordinary people who have trusted them to honour long-term financial promises.

It is remarkable, given these obstacles, that any infrastructure investment is done at all. EDF has invested around £2bn of its share-holders' money in a project that is still not assured of completion. Hitachi, through the Horizon project, has sunk the best part of £1bn, without being confident that it can ever find the billions more to pay for the hoped-for construction of the ABWR at Wylfa in Anglesey.

The UK has a poor infrastructure record in recent decades. This is because capital spending is the easiest thing to cut when government budgets are tight (which includes the present time), and because the UK has been able to live off its outstanding but now crumbling inheritance, chiefly the Victorian investments in sewerage and railways. But the UK is a crowded island with a democratic government, meaning there are many potential vetoes.

The other problem of democracy is said to be the short time horizon, the five-year electoral cycle that inhibits any long-term planning. That seems a fair comment in the case of the multi-decade fiasco of London airports planning. The Roskill Commission of 1971 recommended a new airport in Oxfordshire. The then government rejected this in favour of Foulness on the Thames estuary. Neither was built and we are still arguing about whether to expand Heathrow. Since at least 1990 Thames Water has been preparing for a new reservoir, also in Oxfordshire, to service London's ever-growing population, but has little prospect of ever building it.

By those dismal standards, the new nuclear process has been a triumph. There has been a remarkable degree of continuity through three governments, all of which have made steady progress towards a goal far beyond their terms of office. Nobody can accuse the politicians or civil servants working on the HPC project of being short-termist.

One thing that has surely become clear in the last decade is that nuclear power really is different. It is too large, complex and capital intensive to be left fully to the market. All infrastructure has a form of government guarantee, in that if it fails the state must pick it up and make it work. That was exactly what happened with British Energy (BE) in 2002, a company in which all nuclear risks and liabilities had supposedly been privatized except the ultimate risk of what happened if the company itself failed. So it is impossible fully to privatize the risk and an illusion to believe that any form of contract can do so. The state has a higher responsibility to keep the lights on and to protect the public from the risk of runaway climate change. How it should do that in practice is difficult and messy and will not be achieved by setting a carbon price floor alone.

The Climate Change Act as a doomsday machine

The Climate Change Act of 2008 has turned out to be quite as effective as its supporters hoped, though the majority of the public seem unaware of the remarkable multi-generational commitment they are now bound to. The act has put real pressure on the government to hit future carbon budgets. If asked Why are you signing a 35-year contract for power at double the current price?, the government can say truthfully that it is currently the only way to avoid breaching the fourth carbon budget. Friends of the Earth and the other green sponsors of the act may wonder at what they have wrought.

Well intentioned though the act is, it may not be the best way for the UK to save the planet. The UK is certainly responsible for a share of the inherited CO_2 in the atmosphere, which will linger there for centuries, whatever we do now. But future emissions depend on how much coal China and India burn. All the UK's decarbonization efforts are trivial in comparison with these countries' decisions. It is puzzling therefore that, having identified the importance of carbon capture and storage (CCS) early on, and the benefits it could provide worldwide, not just in the UK, the government has done so little to invest in its development. The UK has some natural advantages here, mainly in having plenty of empty North Sea gas reservoirs, probably suitable for concentrated storage of CO_2.

In addition, the UK's many excellent universities, while already doing a lot in energy research and development, might have achieved a great deal more if they had access to funds equivalent to a fraction of the cost of the HPC project.

Energy policy is always a compromise between affordability and security, with the additional goal in recent decades of cutting pollution. No single energy source currently hits all the targets, so a portfolio approach is needed. Affordability became an important political reality in 2013 when high energy bills coincided with falling incomes to make a lot of British people very angry. The future cost of low-carbon power may yet undermine the consensus and cause a reaction to the Climate Change Act. British governments in reality cannot bind their successors, other than by signing long-term contracts.

Even the promising fall in solar costs won't solve British energy problems, for the obvious reason that it is not sunny at night. Wind too is intermittent. During those inevitable dreary November days when the UK has grey skies and no wind, it will be thermal power, whether gas-fired or nuclear, which keeps the UK moving, lit and warm. Nuclear therefore has a place in the mix for the foreseeable future. It is the longest of long time-horizon investments and the UK has spent nearly a decade thinking about it and building a policy machine to bring it about. HPC and its successors may one day look like foolish and costly commitments that our grandchildren will be paying for. But they should also be reliable sources of low-carbon power that avoid dependence on foreign gas and which offer heat and light on a cold, still winter's night.

REFERENCES

Clean Technica (2014). "India eyes $100 billion investment in renewable energy". 9 November. http://cleantechnica.com/2014/11/09/india-eyes-100-billion-investment-renewable-energy

DECC (2011). *National Policy Statement for Nuclear Power Generation (EN-6)*. Volume I. https://www.gov.uk/government/uploads/system/uploads/attachment_data/file/47859/2009-nps-for-nuclear-volumeI.pdf

IEA (International Energy Agency) (2015). *Global energy-related emissions of carbon dioxide stalled in 2014*. 15 March. http://www.iea.org/news roomandevents/news/2015/march/global-energy-related-emissions-of-carbon-dioxide-stalled-in-2014.html

IEEFA (Institute for Energy Economics and Financial Analysis) (2015). *Briefing Note: Global energy markets in transition*. January. http://www.

ieefa.org/wp-content/uploads/2015/01/IEEFA-BRIEFING-NOTE-Global-Energy-Markets-in-Transition.pdf

Pile, S. (1979). *Book of Heroic Failures: Official handbook of the not terribly good club of Great Britain*. London: Futura.

Appendix:
Nuclear energy

Nuclear engineers sometimes like to describe reactors as big kettles: they are just a source of heat for creating steam to turn a turbine that generates electricity. This is true to the extent that nuclear power is one member of the family of steam-driven power generation, as shown in Figure A.1.

Figure A.1: Different types of electrical energy generation

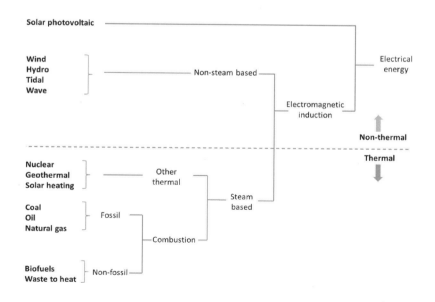

Source: Author

Nuclear fission

Some elements are *fissile* – capable of sustaining a reaction of nuclear fission, in which atoms split into new atoms of a different element, releasing heat. This occurs naturally but it can be made to happen artificially, usually with uranium and plutonium. If it happens rapidly, creating an unstable chain reaction, it can create an explosion, which is the principle of the atomic bomb. Nuclear fission under controlled conditions can produce a steady flow of heat, which is the principle of nuclear electricity stations. The amount of heat produced per weight of fuel is very much larger than in combustion, which is the main advantage of nuclear energy. The main disadvantage is the need to shield the surrounding area from the radiation generated by the reaction and dispose of the dangerous waste afterwards.

A reactor needs a method of control and a coolant to stop it overheating. Control is achieved by introducing neutron-absorbing rods into the reactor, which halt the fission. The rods are made of elements such as boron.

A coolant can be any suitable fluid. Early British reactors used gas, but most modern reactors use water. Other coolants, such as sodium (liquid at very high temperatures), are possible. The choice is a matter of efficiency and cost.

A nuclear reactor is a device to generate heat. To make it into a power station requires a mechanism to take the heat away for use in turning a turbine: ordinary water is used for this. The water is heated in a tube that passes directly through the reactor (one-loop system) or through a heat exchanger that is heated by tube from the reactor (two-loop system). The heated water turns to steam and turns a turbine, just like a conventional gas or coal power station.

The choice of one- or two-loop system is again a matter of efficiency and economics. The dominant reactor type in the world is the pressurized-water reactor (PWR), which uses two loops, as does the Canadian CANDU system. The competing boiling-water reactor (BWR) uses one.

Generation II and III reactors

The first generation of reactors were the original prototypes and research designs. After the Chernobyl disaster of 1986 damaged public confidence in nuclear power, engineers in several countries started to design reactors that were inherently safer. The designs actually built

APPENDIX: NUCLEAR ENERGY **231**

up until the 1990s were informally designated Generation II. The designs built in the West have an excellent safety record and even at the Three Mile Island accident in the USA in 1979 the safety design worked as intended, preventing any significant release of radioactivity to the atmosphere. The 2011 Fukushima disaster resulted from mistakes in the siting and design of the reactor and a poor safety culture, in which Japan had diverged from international best practice. Most other countries with nuclear power don't face earthquake or tsunami risks.

Generation III reactors are both more efficient (more heat and less waste per unit of fuel) and safer. Though there is no official definition of Generation III, all the reactor designs on the market now describe themselves this way. The main reactor types relevant to the UK new build options are shown in Figure A.2.

All the reactors use uranium fuel, enriched to less than 5% and use light water as a coolant. (Light water means "normal" water, as opposed to the "heavy" water (ie with an extra neutron) used in the Canadian CANDU reactors.) They all have very low probabilities of fault and compete mainly on cost.

Figure A.2: Main reactor types competing for UK new build programme

	ABWR	**VVER-1200**	**AP 1000**	**EPR**
Reactor type	BWR	PWR	PWR	PWR
Coolant	Light Water	Light Water	Light Water	Light Water
Designer	GE-Hitachi	Gidropress (*)	Westinghouse	Areva
Electric power, gross MWe	1,420	1,170	1,200	1,770
Net plant efficiency, %	34.4%	33.9%	32%	36%
Plant design life, years	60	60	60	60

Source: Shwageraus, E. (2013) Presentation to Cambridge University Energy Network.
2 December

Glossary

ABWR. Advanced boiling-water reactor.

AGR. Advanced gas-cooled reactor. Higher-performance development of the Magnox reactor.

AP1000. Reactor design by Westinghouse. AP stands for "advanced passive" safety, 1 GW capacity.

Areva. French nuclear company which developed the EPR.

BE. British Energy. Public company owning AGRs and PWR, privatised in 1996 and bought by EDF in 2009

Baseload. That part of electricity demand which is constant over time (in contrast to intermittent demand peaks).

BERR. Department of Business, Enterprise and Regulatory Reform. Created 28 June 2007 after DTI abolished. Responsible for energy policy until the creation of DECC in October 2008. Replaced on 6 June 2009 by BIS.

BIS. Department of Business, Innovation and Skills.

BNFL. British Nuclear Fuels Limited. Government-owned holding company for various nuclear activities, including MOX plant.

Breeder reactor. A reactor that "breeds" its own fuel by converting otherwise useless material such as U-238 into plutonium.

BWR. Boiling-water reactor. Less common form of light-water reactor, using water as coolant and moderator. Differs from PWR in that the steam driving the turbines is generated in the reactor core rather than by separate heat exchangers, so there is a single coolant loop rather than two.

Calder Hall. The UK's first nuclear power station, opened in Cumbria in 1956.

CANDU. Canadian Deuterium Uranium reactor. Heavy-water-moderated reactor from Canada that uses natural uranium fuel.

CCC. Committee on Climate Change. Set up as part of the Climate Change Act to advise government and report on progress towards carbon targets.

CCS. Carbon capture and storage. Technologies to reduce carbon emissions from fossil fuel power stations by removing CO_2 and storing it.

CEGB. Central Electricity Generating Board. Government organization which owned and managed power stations and the national grid in England and Wales in 1957-90.

Centrica. Major UK energy supply company, trading under brand name British Gas.

CfD. Contract for difference. Financial contract that has the effect of fixing the price of a commodity such as power, whose market price fluctuates.

CGN. China General Nuclear Power Group (formerly China Guangdong Nuclear Power Corporation). State-owned Chinese nuclear company based in Shenzhen, Guangdong.

Climate Change Act. Law passed in 2008 committing UK to 80% cut in carbon emissions by 2050.

CNNC. China National Nuclear Corporation. State-owned Chinese nuclear company based in Beijing.

Coolant. Material used to cool nuclear reaction and transfer heat for the generation of steam to turn a turbine. Water (both light and heavy) is the most common coolant. The UK reactors mostly use gas. Sodium is also used in some experimental reactors.

CoRWM. Committee on Radioactive Waste Management. Provides independent advice to the government.

CPR1000. Chinese Generation II reactor developed from the French PWR design.

DECC. Department of Energy and Climate Change.

DEFRA. Department for Environment, Food and Rural Affairs.

Drax. Location of UK's largest coal power station at 4,000 MW, in Yorkshire. Owned by Drax Group plc.

DTI. Department of Trade and Industry. Responsible for energy policy until replaced by BERR in 2007.

EDF. Eléctricité de France. Partially privatized electric utility that owns most of the French electricity industry plus several companies abroad, including in the UK.

EMR. Electricity market reform. Policy of amending the electricity system to include long-term price contracts and capacity markets.

Enrichment. Process of increasing the proportion of U-235 in uranium above its naturally occurring level of 0.7%.

E.ON. Large utility company based in Düsseldorf, Germany.

EPR. European pressurized-water reactor or Evolutionary Power Reactor. Advanced version of PWR with additional safety and efficiency improvements developed by Areva and Siemens, now owned outright by Areva.

ETS. European Emissions Trading System.

Fissile. Having capability of sustaining a chain reaction of nuclear fission. Examples are U-235 and plutonium.

Fission. Process whereby a neutron causes a uranium atom to split into other lighter atoms, releasing energy.

FPL. Florida Power and Light. Florida-based utility company.

GDA. Generic design approval. Process of giving regulatory approval to reactor types.

GDF Suez. Major French utility company.

GE. General Electric. Major US conglomerate, which developed BWR.

Gearing. Ratio of debt to shareholder equity in a company's capital structure.

Generation II and III. Loosely defined categories of nuclear technology, where II refers to the stations built roughly 1990-2010 and III refers to newer designs, particularly those with passive or advanced safety, which are under construction or planned for the future, such as EPR, ABRW and AP1000.

GW. Gigawatt. 1GW = 1,000 megawatts (1000 MW).

Half-life. Period over which radioactivity decays to half its original amount. A measure of the time over which material remains radioactive and potentially dangerous.

Heavy water. Water the molecules of which contain the heavier isotope of hydrogen, deuterium, which has a neutron, whereas ordinary hydrogen has none. Like ordinary ("light") water, heavy water is a good moderator. But heavy water absorbs far fewer neutrons and so allows the use of natural uranium, unlike light water, which needs enriched uranium.

Hinkley Point. Location in Somerset, west England, of existing AGR nuclear station ("B"), proposed new EPR station ("C") and closed Magnox station ("A").

Hitachi. Major Japanese conglomerate, owner with partner GE of the ABWR.

HSE. Health and Safety Executive. Government agency responsible for workplace health and safety, formerly responsible for nuclear safety until creation of separate ONR.

IAEA. International Atomic Energy Agency.

IPCC. Intergovernmental Panel on Climate Change. Scientific body that assesses climate change.

Isotopes. Forms of the same element differing only in the number of neutrons. Isotopes are chemically similar and can only be separated with difficulty, as in the enrichment of uranium to separate out the fissile U-235 isotope from the non-fissile U-238.

IUK. Infrastructure UK. Unit within HM Treasury responsible for promoting infrastructure investment.

kWh. One thousandth of a MWh. Amount of energy consumed by a domestic one-bar electric fire over one hour. Domestic electricity is usually metered in units of 1 kWh.

Leverage. Ratio of debt to shareholder equity in a company's capital structure.

Light water. Ordinary water, as opposed to heavy water.

Load factor. Percentage capacity usage over a year; good performance would be 85-90% for a nuclear station and 30% for onshore wind or solar (because of intermittency).

LWR. Light-water reactor. Reactor using ordinary water as coolant. There are two forms, the PWR and BWR.

Magnox. First-generation British reactor type, named after the magnesium oxide fuel casing.

Moderator. Material used to reduce the speed of neutrons in order to increase the probability of fission. Water is most commonly used but older UK reactors use graphite.

MOX. Mixed oxide fuel. A form of fuel for nuclear reactors which mixes uranium with plutonium recycled from earlier nuclear waste.

MWh. Megawatt hour. Unit of energy calculated from power rating in megawatts, multiplied by number of hours the power rate is sustained.

National grid. High-voltage transmission network in England and Wales, privatized in 1990, owned by the company National Grid Group plc.

National Power. Larger of the two privatized generation companies created out of the CEGB in 1989. Separated in 2000 into Innogy (now owned by German utility RWE) and International Power.

NDA. Nuclear Decommissioning Authority.

NE. Nuclear Electric plc. Government-owned company that owned nuclear power stations in England and Wales until the creation of British Energy and Magnox Electric in 1995.

NII. Nuclear Installations Inspectorate. Nuclear safety regulator succeeded by the ONR.

NuGen. Joint venture company between Toshiba and GDF-Suez, created to build nuclear station at Moorside, Cumbria.

Ofgem. Office of Gas and Electricity Markets, created in 1999 by combining OFFER and the gas regulator.

OND. Office for Nuclear Development. Unit of government set up to facilitate new nuclear power.

ONR. Office for Nuclear Regulation. Government agency responsible for nuclear safety and security across the UK.

PIU. Performance and Innovation Unit. A team within the Prime Minister's office.

Plutonium. Metallic radioactive element. Nuclear weapons and reactors normally use the isotope Pu-239, which has a half-life of about 24,000 years.

PowerGen. Smaller of the two privatized generation companies created out of the CEGB in 1989. Bought by German utility E.ON in 2001.

PWR. Pressurised-water reactor. The most common form of light-water reactor, using ordinary water as coolant and moderator, and slightly enriched uranium as fuel.

ROC. Renewables Obligation Certificate. Mechanism formerly used to ensure electricity suppliers procured renewable energy.

Rosatom. Russian nuclear power company.

RWE. Large international utility company based in Essen, Germany.

Sellafield. Site of BNFL reprocessing operation in Cumbria, north-west England.

Siemens. Major German conglomerate, which co-developed the EPR in partnership with Areva but later sold its interest.

Sizewell. Location in Suffolk, east of England, of existing PWR station ("B"), proposed EPR station ("C") and closed Magnox ("A").

SNL. Scottish Nuclear Limited. Government-owned company that owned nuclear power stations in Scotland from 1989 until merger into British Energy in 1995.

SNPTC. State Nuclear Power Technology Corporation. Chinese state-owned nuclear company.

Toshiba. Major Japanese conglomerate, owner of Westinghouse.

TWh. Terawatt hour. One million MWh, 1,000 GWh.

UKAEA. UK Atomic Energy Authority. Set up in 1954.

UNSCEAR. United Nations Scientific Committee on the Effects of Atomic Radiation

Uranium. Metallic element that is radioactive. Naturally occurring uranium contains 99.3% non-fissile U-238 isotope, with 0.7% fissile isotope U-235.

WANO. World Association of Nuclear Operators. Set up in wake of Chernobyl disaster to foster improved operations and safety in nuclear power.

Westinghouse. Electrical engineering company that developed the AP1000 reactor, originally US but later sold to BNFL and then to Toshiba.

Windscale. Former name of the Sellafield BNFL site and location of air-cooled nuclear reactors used for first military supplies of plutonium in the 1950s.

Index

Also published by UIT Cambridge

Sustainable Energy - without the hot air
David JC MacKay

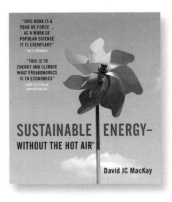

If you've ever wondered how much energy we use, and where it comes from – and where it could come from – but are fed up with all the hot air and "greenwash", this is the book for you. This book cuts through all the contradictory statements from the media, government, and lobbies of all sides. It gives you the numbers and the facts you need, in bite-size chinks, so you can understand the issues yourself.

Sustainable Materials – without the hot air
Julian Allwood and Jonathan Cullen

"*An excellent book … the message is clear and convincing.*" – Bill Gates
This optimistic and richly-informed book evaluates all the options and explains how we can greatly reduce the amount of material demanded and used in manufacturing, while still meeting everyone's needs.

Nuclear 2.0: Why a green future needs nuclear power
Mark Lynas

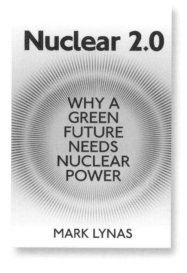

This book is a call for all those who want to see a low-carbon future to join forces and advocate a huge, Apollo-Program-scale investment in wind, solar and nuclear power. Looking at the arguments for and against nuclear power, Mark Lynas lays out a case for why nuclear power is necessary as part of our green future.

Energy and Carbon Emissions
Nicola Terry

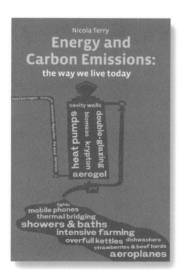

This is a sourcebook of facts and figures about carbon emissions and energy use in the UK. The book emphasizes the impact of domestic consumption and personal lifestyle choices. It's not about politics or ethics: there are no grand designs for how the future should look. Instead, it provides the information you need to calculate not only how you can reduce your carbon footprint, but also how you can save money by reducing your energy bills.